California

Discovery Education | SCIENCE TECHBOOK

The Living Earth

Discovery EDUCATION

Copyright © 2019 by Discovery Education, Inc. All rights reserved. No part of this work may be reproduced, distributed, or transmitted in any form or by any means, or stored in a retrieval or catabase system, without the prior written permission of Discovery Education, Inc.

NGSS is a registered trademark of Achieve. Neither Achieve nor the lead states and partners that developed the Next Generation Science Standards were involved in the production of this product, and do not endorse it.

To obtain permission(s) or for inquiries, submit a request to:
Discovery Education, Inc.
4350 Congress Street, Suite 700
Charlotte, NC 28209
800-323-9084
Education_Info@DiscoveryEd.com

ISBN 13: 978-1-68220-652-2

Printed in the United States of America.

2 3 4 5 6 7 8 9 10 WAL 23 22 21 20 19   B

Acknowledgments
Acknowledgement is given to photographers, artists, and agents for permission to feature their copyrighted material.

Cover and inside cover art: paffy / Shutterstock.com

Table of Contents

How to Use Science Techbook   v

UNIT 1 | Ecosystem Interactions

Concept 1.1 Earth's Spheres   1
Concept 1.2 Ecosystems   13
Concept 1.3 Describing Populations   26
Concept 1.4 Biomes   37

UNIT 2 | Energy, Matter, and Life

Concept 2.1 Energy for Life   50
Concept 2.2 Photosynthesis   62
Concept 2.3 Cellular Respiration   74
Concept 2.4 Nutrient Cycles   85
Concept 2.5 Mechanical and Chemical Weathering   96
Concept 2.6 Erosion and Deposition   107

UNIT 3 | Evolving Earth

Concept 3.1 The History of Life on Earth   118
Concept 3.2 The Development of Earth   129
Concept 3.3 Evidence for Plate Tectonics   139
Concept 3.4 Relative Dating   150
Concept 3.5 Evidence for Evolution   162
Concept 3.6 Mechanisms for Evolution   173

UNIT 4 | Inheritance and Variation

Concept 4.1 Genetics   185
Concept 4.2 DNA   198
Concept 4.3 Transcription and Translation   209
Concept 4.4 Genetic Disorders and Technology   222

UNIT 5 | Cells to Organisms

Concept 5.1 The Chemistry of Life	233
Concept 5.2 Cell Structure and Function	245
Concept 5.3 Cell Transport	256
Concept 5.4 Cell Division	266
Concept 5.5 Asexual and Sexual Reproduction	277
Concept 5.6 Biological Organization and Control	288

UNIT 6 | Human Activities and the Biosphere

Concept 6.1 Natural Resources	300
Concept 6.2 Relationships Between Human Activity and Earth's Systems	311
Concept 6.3 Understanding Climate and Climate Change	322
Concept 6.4 Impacts on Biodiversity	334

Glossary — 346

Course Structure

Science Techbook is a comprehensive teaching and learning package, featuring an award-winning digital platform, a print and digital Student Edition, and a print Teacher's Edition.

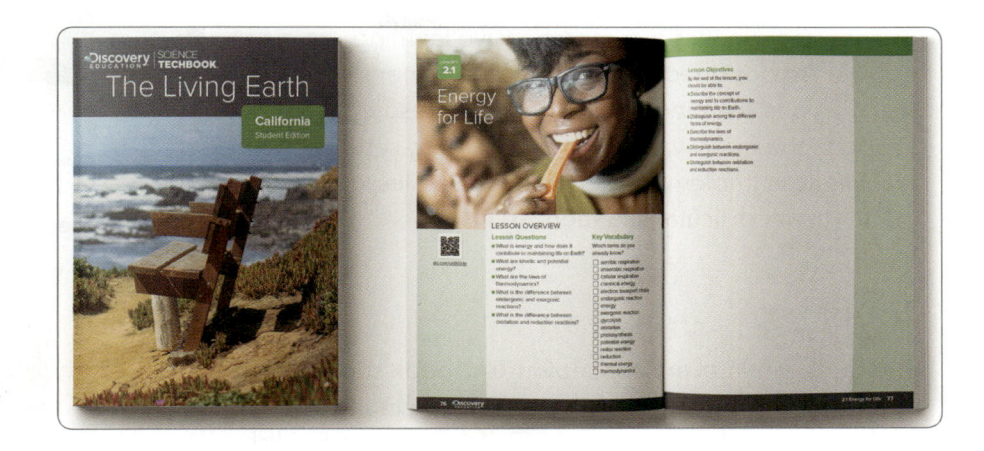

Units

Courses are organized into five to six units. Each unit launches with engaging, real-world anchor phenomena to hook students and to inspire them to ask the questions they want to investigate. At the end of the learning progression, students have the opportunity to solve problems related to the anchor phenomena with the culminating Unit Project. Teachers can "start with the end in mind" by reviewing the 3-dimensional performance-based assessment. This assessment is available in English and Spanish, and helps to prepare students for high-stakes testing.

Concepts

Each unit contains two to five concepts, which are the heart of the learning process. The concept supports the anchor phenomena with the development of student understanding of performance expectations through the use of text, multimedia, Hands-On Activities, and STEM projects.

Every concept . . .

- launches with Investigative Phenomena and a related Can You Explain? Question
- provides multiple pathways for students to demonstrate their learning, including the creation of scientific explanations in the claim, evidence, reasoning format
- encourages STEM career exploration
- helps students summarize their understanding through a required STEM project

The 5Es

Each concept is organized around the research-based 5E Instructional Model: Engage, Explore, Explain, Elaborate with STEM, and Evaluate.

Science Techbook Features

Science Techbook is a comprehensive learning experience designed to support three-dimensional teaching and learning. This flexibility of resources supports the many variations of classroom settings, so teachers can implement standards-based lessons no matter their particular situation.

Tools

The tools within every concept in Techbook support differentiation and cater to the different learning preferences of diverse learners. Switch between English and Spanish, adjust text levels, review student progress on the dashboard, and access the Interactive Glossary for ELD and Vocabulary support.

Core Interactive Text Features

In the digital Core Interactive Text (CIT), students and teachers can have text read aloud, highlight important information, or annotate content with sticky notes. Select any concept text and a reader tool will appear. In addition, content is available in English and authentically translated Spanish, and content can be translated into approximately 90 additional languages using browser translate extensions.

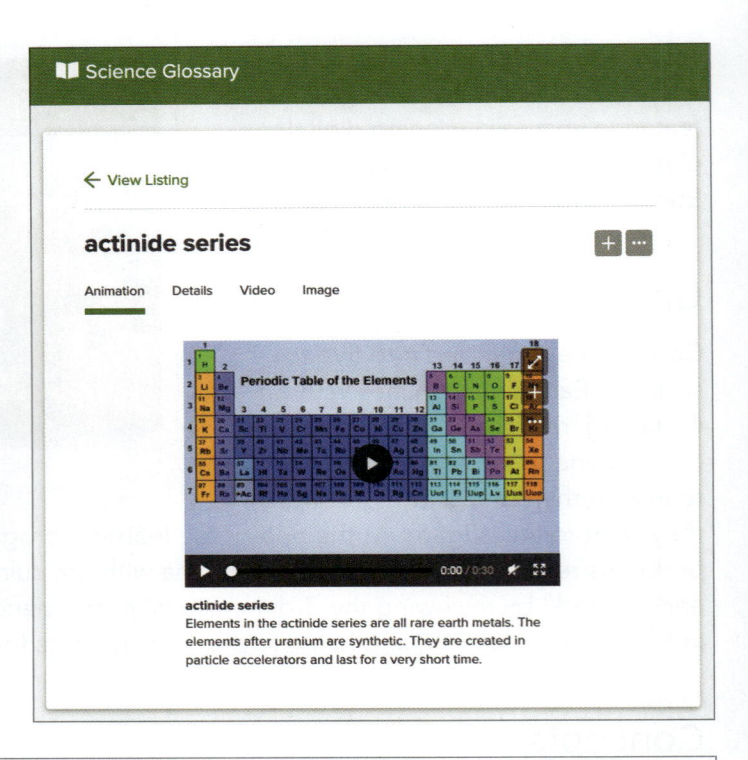

Digital Teacher Materials

In digital Science Techbook, teachers can easily see the student view of content, but also can access additional support using the Teacher Presentation Mode toggle. Teacher notes within each E tab, answer keys to online assessments, additional assessments, and Teacher Guides to Hands-on Activities are visible to only teachers. More in-depth support can be found within each concept's Model Lesson.

Earth's Spheres

The Five E Instructional Model
Science Techbook follows the 5E instructional model. This Model Lesson includes strategies for each of the 5Es. As you design the inquiry-based learning experience for students, be sure to collect data during instruction to drive your instructional decisions. Point-of-use teacher notes are also provided within each E-tab.

Engage   45–90 minutes

Engage Media Resources
The resources found in Engage are intended to stimulate students by exposing them to a phenomenon relevant to the content of the lesson. Engage also provides examples of relevant real-world applications that allow students to begin to make observations and relate the science content to their everyday lives. The Core Interactive Text (CIT) and media resources are carefully designed to prompt students to begin asking questions that they can investigate during the Explore phase of the lesson. They should also start collecting evidence to address the Explain question located at the bottom of the Engage page.

> **TEACHER NOTE** **Investigative Phenomenon:** The investigative phenomenon for this concept addresses what happens when systems are disrupted by severe weather, in this example, by Hurricane Sandy. Use the first paragraph of Engage to elicit student ideas about what a system is. Use the idea of an electrical system in the house to get students thinking about parts of system—appliances, switches, wires—and how they are interrelated and interdependent. Is the electrical system in their home connected to a larger electrical system (the neighborhood electrical sub-station and then the grid) which in turn is connected to other systems with different functions? What happens when different parts of this system are impacted by severe weather conditions? How do other very large external systems such as the atmosphere or oceans impact human constructed systems? Encourage students to recall and share their experiences of such events.

- Core Interactive Text: Exploring Earth's Spheres
- Video: Devastation along the Jersey Seaboard
- Image: Hurricane Sandy Makes Landfall
- Image: System Disruption
- Video: Earth Systems and City Systems Disrupted

Explain Question

The Explain question focuses students on gathering information in the Explore section. The Explain question can be used to

- Record what students already know related to the Explain question.
- Serve as a template or model for students to generate their own scientific questions.
- Collect evidence as students work through the lesson.
- Allow students to reflect on their growth before and after the lesson.

Explain

How do Earth's spheres interact as systems, both within each sphere and with other spheres?

- Image: Spheres within Spheres

Engage Formative Assessment

Technology Enhanced Items (TEIs) found on the Engage page enable you to collect data on students' prior knowledge and identify the common misconceptions they may possess that are related to the topic of study. These items are designed as quick checks for understanding and allow each student one attempt at each question. You can use the data collected to decide whether to assign additional resources to the class, or determine what individual or groups of students may need reinforcement or accelerated learning, prior to completing the Explore portion of the lesson.

TEACHER NOTE Use student responses as a diagnostic assessment to evaluate prior knowledge of the concept. The Model Lesson provides information on common student misconceptions.

TEI Earth Interactions

Before You Begin

What Do I Already Know about Earth's Spheres?

TEACHER NOTE This activity serves as a formative assessment to provide the teacher with feedback on prior knowledge of this topic. Its main focus is to determine student familiarity with changes of state, specifically vaporization/condensation and freezing/melting, which are important processes in Earth's hydrosphere and atmosphere. It is best used as an assignment for students to complete independently.

TEI Energy and Changes of State

TEACHER NOTE This activity may be used as formative assessment to determine student familiarity with the rock cycle, and to identify misconceptions. It is best used as an activity for individuals, followed by a whole-class discussion to air and correct misconceptions.

TEI The Rock Cycle

TEACHER NOTE This open-response formative assessment will demonstrate what students understand from their earlier work with the water cycle. This activity lends itself to a whole-class discussion. Provide students a few minutes to write an initial response, then discuss what they know, or think they know, as a class. During the discussion, pay attention for phrases and statements that suggest these common misconceptions, and elicit discussion that corrects them.

TEI The Water Cycle

TEACHER NOTE This formative assessment activity probes for student knowledge of, and misconceptions about, Earth's magnetic field. It may be assigned to individuals or to student pairs, with a follow up class discussion.

TEI Earth's Magnetic Field

TEACHER NOTE Students should be familiar with Earth's interior layers. This formative assessment probes their understanding of the interactions between Earth's crust and its water, atmosphere, and life. This assessment can be used for group discussion. The teacher can list the interactions that students or groups identify on a white board, black board, or digital projection system.

TEI Earth's Layers

- Video: The Three Phases of Matter
- Video: The Rock Cycle
- Image: The Hydrologic Cycle
- Video: Earth's Geography
- Image: Earth's Interior Structure
- Image: Convection in Earth's Interior

Explore `90 minutes`

Lesson Questions (LQs):

1. What are the major components of the Earth system?

Effective science instruction involves a student-centered rather than a teacher-centered approach. This can be accomplished either with Directed Inquiry or Guided Inquiry, depending on the needs and abilities of your class. Encourage students to select a variety of resources in their pursuit of answers as they work through Explore, with the end goal of constructing their scientific explanation in the Explain tab.

Directed Inquiry	Guided Inquiry
In Directed Inquiry, teachers provide students with a sequence of specified resources, challenging questions, and clear outcomes. Within this context students are given the opportunity to interact independently with each resource as prescribed by the teacher. Often different students groups can be guided through several different resources at the same time. For example, one group could work on a reading passage while a second group conducts a small-group Hands-On Activity with the teacher, and a third group is independently engaged with an online interactive resource.	In Guided Inquiry, students have independence to decide the scope and sequence of their investigations. Using resources from Techbook, students determine for themselves which resources they will Explore to answer the Lesson Questions. It is important to note that each student will choose multiple resources, but no one student is expected to use all the resources available. Students also determine the order in which to explore these resources and how to record their findings.

NGSS Components

SEP	CCC
■ Engaging in Argument from Evidence	■ Structure and Function ■ Stability and Change

Lesson Question: What are the Major Components of the Earth System? `Recommended 90 minutes`

■ Core Interactive Text: What are the Major Components of the Earth System?

TEACHER NOTE Crosscutting Concepts: System and System Models: Throughout this concept, students will learn about the various spheres that make up Earth and how each of them contributes to Earth's ability to sustain life. Students will investigate the system composed of Earth's spheres by creating a model and using it to predict the behavior of this system and what might happen if one of the spheres in the system were significantly altered. This model can be a diagram or a physical model, depending on time, availability of materials, and students' skill levels.

- Video: Introducing Earth's Spheres
- Image: Mapping Earth's Crust
- Video: Introducing the Geosphere
- Video: What Is the Geosphere?
- Image: Inside Earth
- Image: The Geothermal Gradient
- Video: The Sun and the Water Cycle
- Video: Hydrosphere
- Image: Cryosphere
- Reading Passage: Growth of the Cryosphere
- Video: Examining the Oceans

Formative Assessment:

Throughout Explore, Technology Enhanced Items (TEIs) are embedded as multi-dimensional formative checks for understanding. You can use the data they provide to

- assign additional support
- extend learning
- design additional learning tasks to clarify student misconceptions

The Explore TEIs provide students with three attempts to demonstrate their proficiency. Scaffolded feedback is provided for each attempt. If a student does not achieve proficiency by the third attempt, a media asset is provided as an additional learning opportunity.

TEACHER NOTE **Science and Engineering Practice: Using Mathematics and Computational Thinking:** In this item, students must create a mathematical model of Earth's hydrosphere to compute what percentage is represented by each of its components. Students must create equations with one variable and use them to arrive at their answers. Extend this item by having students calculate the total percentage of Earth's water available for human consumption (that is, the percentage found in groundwater and reservoirs, such as lakes and rivers). Then have students discuss the need for water conservation based on their results.

TEI Earth's Water

- Video: Freshwater
- Exploration: Earth's Spheres

TEACHER NOTE Ask students if they are familiar with the hole in the ozone layer and what they think caused it. You could frame this as a murder mystery, with different suspects, including volcanoes, undersea methane vents, cattle ranches, and aerosol sprays, as the possible offenders. The students will play detective, gathering clues and eventually making an "accusation." Clues should come from research, using reliable sources (you may wish to provide students with a list of sources prior to the activity). Students might be surprised to learn that while human activity has certainly had an impact on the atmosphere, natural events, like volcanic eruptions, have had impacts that were as strong if not stronger at the time.

- Video: Introducing the Fluid Sphere
- Video: Atmosphere

TEACHER NOTE **Science and Engineering Practice: Analyzing and Interpreting Data:** In this item, students will analyze the data in the sample charts in order to make valid and reliable scientific claims about the potential effects of the changes shown in each chart on global temperature. Ask students what they know about greenhouse gases and their impact on global climate. Point out that while carbon dioxide may get the most attention, there are other greenhouse gases, including water vapor. Tell students that they will be examining several different chemical composition charts and using that data to determine the effect each change will have on global temperature.

TEI Greenhouse Effect

- Image: Life on Earth
- Video: The Importance of Biodiversity
- Image: Earth's Magnetosphere
- Image: Earth's Magnetic Field
- Hands-On Lab: Magnetic Polarity Sequences
- Video: The Origin and Dynamic Nature of Earth's Magnetic Field

TEACHER NOTE **Crosscutting Concepts: System and System Models:** In this item, students will test their understanding of the interrelationships among the different spheres and how each of the other spheres affects the biosphere's ability to support life by defining each sphere's boundaries and initial conditions, as well as its inputs and outputs. Ask students to think about the interrelationships among the spheres and about how each of them contributes to the complex system that allows life to happen. Ask students what they think would happen if any one of the spheres were to disappear. Have them write down their thoughts and then share some of the responses with the class.

TEI **Beyond the Biosphere**

Explore More Resources

Resources in Explore More Resources support differentiation within your classroom by

- providing additional visualization of content
- affording extension of content to those students ready for acceleration
- offering Lexile reading levels for reading passages

Online explorations and hands-on experiences are provided so that students can conduct virtual investigations, collect and design investigations, and collect and analyze data; these skills are essential to developing scientific understanding.

Explain `45–90 minutes`

In Explore, students
1. uncovered scientific understandings
2. conducted investigations
3. analyzed data, text, and other media resources
4. collected evidence to support their scientific explanation

In Explain, provide students with time to formally compose their scientific explanations around the CYE or student-generated questions using evidence collected from Explore.

Scientific explanations are student responses, either written or orally presented, that explain scientific phenomena based upon evidence. Developing a scientific explanation requires students to analyze and interpret data to construct meaning out of the data. There are three main components to the scientific explanation: the claim, the evidence, and the reasoning.

To help students to communicate their scientific explanations, allow them to utilize the multimedia creation tools such as Board Builder and Whiteboard. Remind them that they may upload image, audio, and video files using the "attach file" option to communicate their scientific explanations.

Students may construct their scientific explanations individually or within a small group of students. Students should communicate their explanations with other classmates, and provide constructive criticism and refine their explanations prior to submission to the teacher. If explanations are used as a formative assessment, you can provide additional feedback and comments to support students as they refine their explanations.

CAN YOU EXPLAIN?

How do Earth's spheres interact as systems both within each sphere and among other spheres?

Elaborate with STEM   45–135 minutes

*Elaborate with STEM are optional extension resources available after students have demonstrated proficiency with standards addressed previously in the concept.

NGSS Components

SEP	CCC
■ Obtaining, Evaluating, and Communicating Information ■ Asking Questions and Defining Problems ■ Planning and Carrying Out Investigation ■ Using Mathematics and Computational Thinking ■ Constructing Explanations and Designing Solutions	■ Cause and Effect ■ Structure and Function

STEM In Action   45 minutes

STEM in Action ties the scientific concepts to real-world applications, with many connecting to STEM careers. Technology Enhanced Items (TEIs) expect students to critically read the Core Interactive Text (CIT) and review the provided media resources.

Applying Earth's Spheres

- Core Interactive Text: Applying Earth's Spheres
- Image: Magnetic Pole Reversal
- Video: Geomagnetism on Earth
- Video: Hidden Oil Plumes
- Image: Pattern of Magnetic Reversals
- Reading Passage: Magnetic Pole Reversal
- Reading Passage: Interview with an Atmospheric Scientist

STEM Project Starter

STEM Project Starters provide additional real-world contexts that require students to apply and extend their content knowledge related to the concept. STEM Project Starters can also serve as an alternative instructional hook presented at the beginning of the learning progression. The project can then be revisited throughout and at the end of the 5E learning cycle, for students to apply content knowledge.

STEM Project Starter: Oil Leak!   `Recommended 45 minutes`

How do the interactions of Earth's spheres affect the spread of oil after a deep-water leak?

> **TEACHER NOTE** This STEM activity may be used as a formative assessment. It allows students to demonstrate their ability to interpret and correlate information displayed in two graphs, while applying their understanding of the interactions of Earth's hydrosphere and geosphere in a real-life situation.

- Image: Side View of Oil Leak, Day 2
- Image: Map View of Oil Leak, Day 5

TEI Oil Leak!

STEM Project Starter: Earth's Temperature: The Big Picture   `Recommended 60 minutes`

Where does temperature change faster: as you drill into Earth or as you rocket away from its surface?

> **TEACHER NOTE** This Exploration offers an opportunity for formative assessment. It combines science and mathematics as students create a master graph of Earth's temperature from its core through its outer atmosphere. They then relate the range of temperature for the geosphere and the atmosphere, and the rate at which the temperature changes.
>
> Students will find it helpful to keep the Exploration open in one window, and the Data/Graphing Tool in another.

- Exploration: Earth's Spheres
- Science Tool: Data/Graphing Tool

TEI Make a Data Table

TEI Make a Graph

TEI Interpret the Graph

STEM Project Starter: Finding the Stripes

`Recommended 90 minutes`

Who helped discover paleomagnetism?

- Video: The Sea Floor Is Spreading
- Video: Magnetic Field Reversal

TEACHER NOTE This is a formative assessment. This STEM project highlights the importance of technological developments to scientific discoveries, in the setting of the historical development of a scientific theory.

This project has potential as a cross-disciplinary extension project with a cooperating teacher. For example, a connection to history can be made by emphasizing the societal context in which the scientists worked. English courses could connect through the practice of epistolary writing. Another option is to tie in with dramatic arts. Instead of having student write letters, have students act as their scientists during a mock panel discussion moderated by the teacher or by an informed classmate (ignoring the fact that not all scientists on the list were alive at the same time).

TEI Letters

STEM Project Starter: Deepwater Horizon Update

`Recommended 90 minutes`

What are the long-term effects of the Deepwater Horizon oil spill on Earth's hydrosphere and biosphere?

TEACHER NOTE This STEM project provides an opportunity for a summative assessment, as it expands student understanding of the interactions of Earth's spheres through research into the long-term status of the 2010 oil spill in the Gulf of Mexico. As students research the tools and techniques used to monitor the situation, they develop an appreciation for the role of technology and engineering in the pursuit of scientific knowledge.

Any generally well-researched, objective, fact-checked source should be considered acceptable. Articles from news outlets, public education sites from government and educational institutions, and interviews with scientists will be more appropriate for most high school students than scientific journals.

As an alternative approach, have students write more detailed and formal reports to include an introduction, supportive paragraphs in the body, and a conclusion. The supportive paragraphs should touch on the effects of the oil spill and the tools used to monitor it, with a minimum of one paragraph each. Have students form pairs and perform a peer review on each other's writing. Students can use this feedback to improve their transitions, language, style, and introduction and conclusion.

TEI Deepwater Horizon

TEI Predicting Interactions

Evaluate `45–90 minutes`

Explain? Question:
How do Earth's spheres interact as systems, both within each sphere and among other spheres?

Lesson Questions (LQ):
- What are the major components of the Earth system?

Throughout instruction and the 5E learning cycle, you will have collected formative assessment data to drive the assignment of resources and experiences to students. Evaluate is intended to include summative assessment checks for proficiency. You can use the CYE and Lesson Questions for the concept as a summative assessment in a variety of ways such as these:

- Post each Lesson Question (LQ) in various locations in the classroom, and have small groups of students generate claim statements related to the Lesson Question (LQ). Other students can add to the claim, or refute the claim, during a gallery walk where they place additional pieces of evidence on each Lesson Question (LQ) poster.
- Assign small groups of students to each Lesson Question (LQ) and have the groups generate a poster, board, graphic, or piece of text that answers the question. Use a jigsaw approach and create a second set of groups that contain members from each Lesson Question (LQ) group to share their ideas.
- Ask students to return to their initial ideas for the Explain question and add additional details and evidence.

Encourage students to review the concept review and complete the Student Self-Check practice assessment prior to assigning the Summative Teacher Concept assessment.

- Student Review and Practice Assessment
- Teacher Concept Assessments

Ecosystems

The Five E Instructional Model

Science Techbook follows the 5E instructional model. This Model Lesson includes strategies for each of the 5Es. As you design the inquiry-based learning experience for students, be sure to collect data during instruction to drive your instructional decisions. Point-of-use teacher notes are also provided within each E-tab.

Engage **45–90 minutes**

Engage Media Resources

The resources found in Engage are intended to stimulate students by exposing them to a phenomenon relevant to the content of the lesson. Engage also provides examples of relevant real-world applications that allow students to begin to make observations and relate the science content to their everyday lives. The Core Interactive Text (CIT) and media resources are carefully designed to prompt students to begin asking questions that they can investigate during the Explore phase of the lesson. They should also start collecting evidence to address the Explain question located at the bottom of the Engage page.

> **TEACHER NOTE** **Investigative Phenomenon:** Students will have learned about ecosystems in earlier grades. Usually, ecosystems are depicted as large-scale "natural" systems such as those found in forests, lakes, or deserts. In fact, the concept of an ecosystem—a community of organisms together with their physical environment—may be applied at any scale and can also be applied to manmade environments. Use the image "Dust Mites" to frame the question, "Do you think that this picture depicts part of an ecosystem?" Have the students work in pairs or small groups to discuss this question and then share and justify their answers. Extend the discussion as to what other organisms may be part of this ecosystem, perhaps what a food chain in this ecosystem could look like (the mites feed on human skin). Also have students consider the abiotic components of this ecosystem.

- Core Interactive Text: Exploring Ecosystems
- Image: Dust Mites
- Video: Plant and Animal Interdependence
- Video: Introduction to Ecology: Ecosystems and Biomes
- Video: Science in Progress: Adaptive Radiation

Explain Question

The Explain question focuses students on gathering information in the Explore section. The Explain question can be used to

- Record what students already know related to the Explain question.
- Serve as a template or model for students to generate their own scientific questions.
- Collect evidence as students work through the lesson.
- Allow students to reflect on their growth before and after the lesson.

Explain

What components are necessary for the function of a successful ecosystem?

- Image: Examining Ecosystems

Engage Formative Assessment

Technology Enhanced Items (TEIs) found on the Engage page enable you to collect data on students' prior knowledge and identify the common misconceptions they may possess that are related to the topic of study. These items are designed as quick checks for understanding and allow each student one attempt at each question. You can use the data collected to decide whether to assign additional resources to the class, or determine what individual or groups of students may need reinforcement or accelerated learning, prior to completing the Explore portion of the lesson.

> **TEACHER NOTE** Use this student response to evaluate students' prior knowledge of the ecosystem concept. The Model Lesson provides information on common student misconceptions. Student answers should indicate that they are aware of the need to acknowledge the smallest organisms, such as bacteria, and the larger ones, such as the trees. They should also include inorganic references to such things as minerals, water, and sunlight. This may be a good question to have students discuss in a small group before responding.

TEI Ecosystem Analysis

Before You Begin

What Do I Already Know about Ecosystems?

> **TEACHER NOTE** This activity is intended to provide the teacher with feedback on prior knowledge of this topic. It is a good activity to use with individual students. However, a small group discussion in advance may help students differentiate between biomes and communities. Students should be aware that the terms biome and community are not synonymous with ecosystem, but represent different levels of biological hierarchy. Consider using this content in a paired activity to help struggling or ESL students learn about basic terms related to ecosystems.

TEI Characteristics of Biomes and Communities

TEACHER NOTE This is a formative activity. This connected response may be used for individuals or small groups that can discuss their reasoning before responding. Students may have the misconception that ecosystems must be permanent, but some ecosystems are temporary. They may also think the ecosystem only involves living things, but nonliving things are part of an ecosystem. Students who show difficulties can be directed back to the Model Lesson.

TEI Characteristics of an Ecosystem

TEACHER NOTE This activity is intended to provide the teacher with feedback on prior knowledge of this topic. Its main focus is to determine whether students understand that the definition of species involves the group being able to interbreed. Just looking similar is not enough to put two categories of living things in the same species group, as in the case of two different birds. Pigeons and eagles are both birds and yet cannot interbreed.

The Model Lesson has additional video segments on this topic if students struggle. This is best answered by individual students.

TEI Definition of Species

- Video: Characteristics of Living Things
- Image: Interbreeding and Dogs
- Video: Flying Fish: Soaring to Safety
- Image: Biome Definition

Explore   180 minutes

Lesson Questions (LQs):

1. What are the characteristics of a population, a community, and an ecosystem?
2. How do organisms interact within a community?
3. How is an ecosystem established?
4. What are the unique characteristics of water communities?
5. What external factors influence ecosystems? How do species, populations, communities, and entire ecosystems respond to these external factors?

Effective science instruction involves a student-centered rather than a teacher-centered approach. This can be accomplished either with Directed Inquiry or Guided Inquiry, depending on the needs and abilities of your class. Encourage students to select a variety of resources in their pursuit of answers as they work through Explore, with the end goal of constructing their scientific explanation in the Explain tab.

Directed Inquiry	Guided Inquiry
In Directed Inquiry, teachers provide students with a sequence of specified resources, challenging questions, and clear outcomes. Within this context students are given the opportunity to interact independently with each resource as prescribed by the teacher. Often different students groups can be guided through several different resources at the same time. For example, one group could work on a reading passage while a second group conducts a small-group Hands-On Activity with the teacher, and a third group is independently engaged with an online interactive resource.	In Guided Inquiry, students have independence to decide the scope and sequence of their investigations. Using resources from Techbook, students determine for themselves which resources they will Explore to answer the Lesson Questions. It is important to note that each student will choose multiple resources, but no one student is expected to use all the resources available. Students also determine the order in which to explore these resources and how to record their findings.

NGSS Components

SEP	CCC
■ Developing and Using Models ■ Analyzing and Interpreting Data ■ Engaging in Argument from Evidence	■ Systems and System Models ■ Structure and Function ■ Stability and Change

Lesson Question: What Are the Characteristics of a Population, a Community, and an Ecosystem?

Recommended 30 minutes

> **TEACHER NOTE Connections: Crosscutting Concept: Systems and System Models:** In this concept, students will understand the structure of ecosystems and the interactions within them. They will use models to simulate the flow of energy, matter, and interactions within systems. Use the "Visual Walkabout" strategy to introduce students to this concept. Images can include an owl, a chipmunk, a single deer, a group of deer, a wolf, a forest ecosystem, and other forest animals or details of the biotic and abiotic factors in a forest ecosystem. Before students begin their gallery walk, inform them that all the images in the gallery are parts of a single system. As students ask questions and make connections between the images, they will be creating a simple model of a forest ecosystem. The "Visual Walkabout" strategy is found on the Professional Learning tab. Click on Strategies & Resources, then click on Spotlight on Strategies (SOS). Now click on Instructional Hook, then click on Spotlight on Strategies: Visual Walkabout.

- Core Interactive Text: What Are the Characteristics of a Population, a Community, and an Ecosystem?
- Image: Population Distribution Types
- Video: Coral Communities of Palau
- Hands-On Activity: English Language Proficiency Activity: Ecosystems
- Video: Biodiversity in Ecosystems
- Video: Aye-Aye: Filling a Niche

Formative Assessment:

Throughout Explore, Technology Enhanced Items (TEIs) are embedded as multi-dimensional formative checks for understanding. You can use the data they provide to

- assign additional support
- extend learning
- design additional learning tasks to clarify student misconceptions

The Explore TEIs provide students with three attempts to demonstrate their proficiency. Scaffolded feedback is provided for each attempt. If a student does not achieve proficiency by the third attempt, a media asset is provided as an additional learning opportunity.

> **TEACHER NOTE Practices: Science and Engineering Practice: Developing and Using Models:** In this item, students classify different interactions between organisms according to models of ecosystem relationships. They develop and use a model based on evidence to illustrate the relationships between systems or between components of a system. To extend this activity, have students work in groups to draw their own concentric circle models, using their school to create an analogy for each level.

TEI Population, Community, Ecosystem

Lesson Question: How Do Organisms Interact Within a Community?

Recommended 30 minutes

- Core Interactive Text: How Do Organisms Interact Within a Community?

TEACHER NOTE Practices: Science and Engineering Practice: Engaging in Argument from Evidence: Give students the opportunity to make and defend claims about relationships in the natural world based on evidence. Provide several examples of ecological relationships, pretending that they are from a scientist's field journal. Have students identify each organism in the relationship and explain whether each organism benefits, is harmed, or neither by citing evidence from the field journal.

- Video: The Pork Tapeworm
- Image: School of Fish
- Video: Meerkats Raising Young
- Video: Energy in Ecosystems
- Video: Food Chains and Food Webs
- Video: Ecological Pyramids

TEACHER NOTE Connections: Crosscutting Concept: Structure and Function: In this item, students investigate ecosystems by examining the properties of interactions between organisms. They examine the function of each organism in these interactions to classify how the relationship functions within the system. Before they complete this item, help students review these relationships by using the ABC Summary strategy to summarize what they have learned in this concept according to the letters of the alphabet. The "ABC Summary" strategy is found on the Professional Learning tab. Click on Strategies & Resources, then click on Spotlight on Strategies (SOS). Now click on Vocabulary Development, then click on Spotlight on Strategies: ABC Summary.

TEI Ecological Relationships

Lesson Question: How Is an Ecosystem Established?

`Recommended 20 minutes`

- Core Interactive Text: How Is an Ecosystem Established?
- Video: Ecological Succession
- Video: A Keystone Species
- Video: Climate Change and Invasive Species

TEACHER NOTE Misconception: Students may believe that an ecosystem is composed of a neat hierarchy of independent biological systems. However, ecosystems often exist within other ecosystems, which may have complex interdependent relationships.

Lesson Question: What Are the Unique Characteristics of Water Communities?

`Recommended 20 minutes`

- Core Interactive Text: What Are the Unique Characteristics of Water Communities?
- Video: Aquatic Biomes

TEACHER NOTE Misconception: Students may believe that aquatic organisms can easily adapt to either saltwater or freshwater. In fact, most aquatic organisms are adapted to their specific environment and are sensitive to changes in salinity.

TEACHER NOTE Practices: Science and Engineering Practice: Analyzing and Interpreting Data: In this item, students analyze and interpret data to understand how aquatic ecosystems change over time. Because students may be unfamiliar with interpreting gradient maps, such as those presented in this item, it may be necessary to explain that darker colors indicate greater salinity. As an extension, provide a large version of the map and have students build a physical model by sprinkling salt on the map, placing more salt in regions that are darker in color. Have students research which organisms live in the freshwater, marine, or estuarine ecosystems, and add pictures of these organisms to their model. Ask students to look for patterns in the relationships among the organisms that are part of their model.

TEI Estuary Salinity

Lesson Question: What External Factors Influence Ecosystems? How Do Species, Populations, Communities, and Entire Ecosystems Respond to These External Factors?

Recommended 80 minutes

- Core Interactive Text: What External Factors Influence Ecosystems? How Do Species, Populations, Communities, and Entire Ecosystems Respond to These External Factors?
- Video: Ecosystems Respond to Disturbances
- Video: Climate Change and Ecosystems
- Video: Rainforest Removal
- Hands-On Lab: Stream Ecology

TEACHER NOTE Connections: Crosscutting Concept: Stability and Change: In this item, students predict how changes in temperature would change the ecosystem. Students understand how an ecosystem is a dynamic system where temperature will change over time. Help students make these predictions with the Myth Bustin'! strategy. Provide students with statements regarding the Great Lakes and have them determine whether the statements are fact or fiction. For example: Forests prefer warm weather over cold weather (fiction). Different trees grow in different climates (fact). The "Myth Bustin'!" strategy is found on the Professional Learning tab. Click on Strategies & Resources, then click on Spotlight on Strategies (SOS). Now click on Inference and prediction, then click on Spotlight on Strategies: Myth Bustin'!

TEI The Changing Great Lakes

Explore More Resources

Resources in Explore More Resources support differentiation within your classroom by

- providing additional visualization of content
- affording extension of content to those students ready for acceleration
- offering Lexile reading levels for reading passages

Online explorations and hands-on experiences are provided so that students can conduct virtual investigations, collect and design investigations, and collect and analyze data; these skills are essential to developing scientific understanding.

Explain `45–90 minutes`

In Explore, students
1. uncovered scientific understandings
2. conducted investigations
3. analyzed data, text, and other media resources
4. collected evidence to support their scientific explanation

In Explain, provide students with time to formally compose their scientific explanations around the CYE or student-generated questions using evidence collected from Explore.

Scientific explanations are student responses, either written or orally presented, that explain scientific phenomena based upon evidence. Developing a scientific explanation requires students to analyze and interpret data to construct meaning out of the data. There are three main components to the scientific explanation: the claim, the evidence, and the reasoning.

To help students to communicate their scientific explanations, allow them to utilize the multimedia creation tools such as Board Builder and Whiteboard. Remind them that they may upload image, audio, and video files using the "attach file" option to communicate their scientific explanations.

Students may construct their scientific explanations individually or within a small group of students. Students should communicate their explanations with other classmates, and provide constructive criticism and refine their explanations prior to submission to the teacher. If explanations are used as a formative assessment, you can provide additional feedback and comments to support students as they refine their explanations.

CAN YOU EXPLAIN?
What components are necessary for the function of a successful ecosystem?

Elaborate with STEM `45–135 minutes`

*Elaborate with STEM are optional extension resources available after students have demonstrated proficiency with standards addressed previously in the concept.

NGSS Components

SEP	CCC
■ Obtaining, Evaluating, and Communicating Information ■ Asking Questions and Defining Problems ■ Planning and Carrying Out Investigation ■ Using Mathematics and Computational Thinking ■ Constructing Explanations and Designing Solutions	■ Cause and Effect ■ Structure and Function

STEM In Action `45 minutes`

STEM in Action ties the scientific concepts to real-world applications, with many connecting to STEM careers. Technology Enhanced Items (TEIs) expect students to critically read the Core Interactive Text (CIT) and review the provided media resources.

Applying Ecosystems

- Core Interactive Text: Applying Ecosystems
- Video: The Threat to Biodiversity
- Image: Clearcut Logging
- Image: Sunflower Phytoremediation
- Video: Six Ways Mushrooms Can Save the World
- Video: Threats to Freshwater Mussels and the Consequences for Ecosystems
- Video: Chernobyl: A Strontium Disaster
- Image: Contaminated Runoff Reaches a River

TEACHER NOTE This is a summative activity that allows students to explore a connection to developing technology and the content of the lesson. This activity can be done in small groups or can be assigned as an individual project.

TEI Bioremediation

STEM Project Starters

STEM Project Starters provide additional real-world contexts that require students to apply and extend their content knowledge related to the concept. STEM Project Starters can also serve as an alternative instructional hook presented at the beginning of the learning progression. The project can then be revisited throughout and at the end of the 5E learning cycle, for students to apply content knowledge.

STEM Project Starter: Summer Swimming   `Recommended 45 minutes`

Swimming in the summer is refreshing. How can high temperatures make some places unsafe for humans to swim?

> **TEACHER NOTE** Students can apply what they have learned about water ecosystems while interpreting this chart about bacteria levels in a freshwater ecosystem. This is an activity you may introduce to the whole class to give the opportunity to review aspects of a water ecosystem. Students may then complete the activity in pairs or independently.
>
> Inform students that they should structure their letters to the mayor to include an introduction, supporting paragraphs, and conclusion. The first paragraph should introduce the problem with the sewage system. Students should include supporting paragraphs in the body of the letter that make use of data provided by the graph. Finally, students should provide a solution (or solutions) to bioremediate the water bacteria problem in the conclusion of their letters. Instruct students to use an active, academic voice rather than a casual or informal tone for their reports.

TEI Summer Swimming

STEM Project Starter: Tertiary or Secondary Consumer?   `Recommended 90 minutes`

This bird eats both seeds and insects. Is its trophic level secondary or tertiary?

> **TEACHER NOTE** After students complete the Exploration entitled Ecosystems, have them complete this project. You may wish to have students discuss their answers with partners before they respond.

■ Exploration: Ecosystems

TEI Relationships in Ecosystems

STEM Project Starter: Red Squirrel Population Changes `Recommended 90 minutes`

What caused the changes in red squirrel populations in a community?

> **TEACHER NOTE** This summative assessment enables students to apply what they have learned about community interactions and to analyze the effects of the introduction of an exotic species into an ecosystem. This project extends student learning about species interactions by providing an example of how competition between an invasive species and a species established in a community can lead to changes in the ecosystem. Prior to beginning this project, it is recommended that students complete the assignment Species Interactions.

> **TEACHER NOTE** Have students work individually on this project. Encourage students to conduct additional research on red and grey squirrel species to provide more background on the differences and similarities between these species. After students have completed the entire project, bring them together as a group and discuss their answers to the questions. Ask them whether the data actually show that the grey squirrels out-competed the red squirrels or whether they think the data are correlational only. Ask them to think about what is required to show that one event actually causes another and is not linked by some other relationship that is not causal in nature. Ask students what type of study would have to be done to directly show that competition with grey squirrels is the cause of the decline in red squirrel populations.

- Video: Types of Interactions within Ecosystems

TEI **Analyzing Population Changes**

Evaluate `45–90 minutes`

Explain Question:
What components are necessary for the function of a successful ecosystem?

Lesson Questions (LQ):
- What are the characteristics of a population, a community, and an ecosystem?
- How do organisms interact within a community?
- How is an ecosystem established?
- What are the unique characteristics of water communities?
- What external factors influence ecosystems? How do species, populations, communities, and entire ecosystems respond to these external factors?

Throughout instruction and the 5E learning cycle, you will have collected formative assessment data to drive the assignment of resources and experiences to students. Evaluate is intended to include summative assessment checks for proficiency. You can use the CYE and Lesson Questions for the concept as a summative assessment in a variety of ways such as these:

- Post each Lesson Question (LQ) in various locations in the classroom, and have small groups of students generate claim statements related to the Lesson Question (LQ). Other students can add to the claim, or refute the claim, during a gallery walk where they place additional pieces of evidence on each Lesson Question (LQ) poster.
- Assign small groups of students to each Lesson Question (LQ) and have the groups generate a poster, board, graphic, or piece of text that answers the question. Use a jigsaw approach and create a second set of groups that contain members from each Lesson Question (LQ) group to share their ideas.
- Ask students to return to their initial ideas for the Explain question and add additional details and evidence.

Encourage students to review the concept review and complete the Student Self-Check practice assessment prior to assigning the Summative Teacher Concept assessment.

- Student Review and Practice Assessment
- Teacher Concept Assessments

Describing Populations

The Five E Instructional Model

Science Techbook follows the 5E instructional model. This Model Lesson includes strategies for each of the 5Es. As you design the inquiry-based learning experience for students, be sure to collect data during instruction to drive your instructional decisions. Point-of-use teacher notes are also provided within each E-tab.

Engage   45–90 minutes

Engage Media Resources

The resources found in Engage are intended to stimulate students by exposing them to a phenomenon relevant to the content of the lesson. Engage also provides examples of relevant real-world applications that allow students to begin to make observations and relate the science content to their everyday lives. The Core Interactive Text (CIT) and media resources are carefully designed to prompt students to begin asking questions that they can investigate during the Explore phase of the lesson. They should also start collecting evidence to address the Explain question located at the bottom of the Engage page.

> **TEACHER NOTE** **Investigative Phenomenon:** The term *population* is often used as synonym for human population. To get students thinking about populations in terms of other organisms, have them view the video Declining Animal Populations. After they have viewed the video, have students discuss in small groups why animal populations may be in decline. Does this apply to all animals, or are there species that they think could be increasing in numbers? Do they think a similar decline exists in plant populations? If so, why? Have students consider why measuring animal populations is difficult. What methods could they suggest to measure animal populations?

- Core Interactive Text: Declining Animal
- Video: Declining Animal Populations
- Video: Science Nation: Emperor Penguins and Climate Change

Explain Question

The Explain question focuses students on gathering information in the Explore section. The Explain question can be used to

- Record what students already know related to the Explain question.
- Serve as a template or model for students to generate their own scientific questions.
- Collect evidence as students work through the lesson.
- Allow students to reflect on their growth before and after the lesson.

Explain
Explain the factors that impact squirrel populations.

■ Image: Nuts for Squirrels

Engage Formative Assessment
Technology Enhanced Items (TEIs) found on the Engage page enable you to collect data on students' prior knowledge and identify the common misconceptions they may possess that are related to the topic of study. These items are designed as quick checks for understanding and allow each student one attempt at each question. You can use the data collected to decide whether to assign additional resources to the class, or determine what individual or groups of students may need reinforcement or accelerated learning, prior to completing the Explore portion of the lesson.

TEACHER NOTE Use this student response to evaluate students' prior knowledge of the concept. The Model Lesson provides information on common student misconceptions. This item should be completed on an individual basis.

TEI Your Ideas

Before You Begin
What Do I Already Know about Describing Populations?

TEACHER NOTE This formative assessment item is intended to provide the teacher with feedback on prior knowledge of this topic. Students should have already learned about abiotic and biotic factors in elementary and middle school. Students could work in small groups to classify the factors and discuss their answers.

TEI Abiotic and Biotic Factors

TEACHER NOTE This formative question could be used to assess a common misconception about populations. If the student does not select C as the correct answer, a discussion can begin about the factors that affect populations. If the student does not selects D as the correct answer, a brainstorming session about ways in which one population can affect another can begin. It would be best to start this as a think-pair-share activity and then discuss it further as a large group.

TEI Populations

TEACHER NOTE This formative activity is intended to provide the teacher with feedback on prior knowledge of this topic and common sources of confusion when a student does not understand the different levels of biological organization.

If students need review of these terms, they can do it now before moving onto more complex concepts. It would work well to have students do the question individually and then discuss the answers as a group. One way to extend this activity is to have students come up with examples of each of these terms.

TEI Levels of Organization

- Video: Science Nation: Catching a Coral Killer
- Video: Tables and Bar Graphs – Drums
- Video: What Is a Population?

Explore   180 minutes

Lesson Questions (LQs):

1. What are the factors of population demographics?
2. What is the difference between density-dependent and density-independent factors?
3. What are the differences between exponential and logistic growth?
4. How is population growth affected by carrying capacity?
5. How are age structures created?
6. What are the differences between type I, type II, and type III survivorship curves?

Effective science instruction involves a student-centered rather than a teacher-centered approach. This can be accomplished either with Directed Inquiry or Guided Inquiry, depending on the needs and abilities of your class. Encourage students to select a variety of resources in their pursuit of answers as they work through Explore, with the end goal of constructing their scientific explanation in the Explain tab.

Directed Inquiry	Guided Inquiry
In Directed Inquiry, teachers provide students with a sequence of specified resources, challenging questions, and clear outcomes. Within this context, students are given the opportunity to interact independently with each resource as prescribed by the teacher. Often different students groups can be guided through several different resources at the same time. For example, one group could work on a reading passage while a second group conducts a small-group Hands-On Activity with the teacher, and a third group is independently engaged with an online interactive resource.	In Guided Inquiry, students have independence to decide the scope and sequence of their investigations. Using resources from Techbook, students determine for themselves which resources they will Explore to answer the Lesson Questions. It is important to note that each student will choose multiple resources, but no one student is expected to use all the resources available. Students also determine the order in which to explore these resources and how to record their findings.

NGSS Components

SEP	CCC
■ Developing and Using Models ■ Constructing Explanations and Designing Solutions	■ Patterns ■ Cause and Effect ■ Systems and System Models ■ Stability and Change

Lesson Question: What Are the Factors of Population Demographics?

`Recommended 30 minutes`

> **TEACHER NOTE Connections: Crosscutting Concept: Stability and Change:** In this concept, students understand how studying population ecology helps us explain how populations change and remain stable. They learn how scientists quantify and model changes in population over short and long periods of time. To introduce this concept to students, share the video segment "What Is a Population?" with students and instruct them to make Frayer maps for the word *population*. Ask them to focus on how a population forms and changes over time. This strategy is found by searching for "Vocabulary Quadrants" on Science Techbook.

- Core Interactive Text: What Are the Factors of Population Demographics?
- Video: What Is a Population?
- Video: Prairie Dog City
- Image: School of Fish
- Image: California Poppies
- Video: Economy and Demographics of Greenland

Formative Assessment:

Throughout Explore, Technology Enhanced Items (TEIs) are embedded as multi-dimensional formative checks for understanding. You can use the data they provide to

- assign additional support
- extend learning
- design additional learning tasks to clarify student misconceptions

The Explore TEIs provide students with three attempts to demonstrate their proficiency. Scaffolded feedback is provided for each attempt. If a student does not achieve proficiency by the third attempt, a media asset is provided as an additional learning opportunity.

> **TEACHER NOTE Connections: Crosscutting Concept: Patterns:** In this item, students observe patterns in population changes and distributions. To extend this activity, students can use Discovery resources and other sources, such as the USDA, to research the honeybee population changes since 2006. Scientists disagree about the extent of the population changes and their potential impact, so students can take sides and cite evidence from their research in a class debate.

TEI Honeybee Populations

Lesson Question: What Is the Difference Between Density-Dependent and Density-Independent Factors?

`Recommended 30 minutes`

- Core Interactive Text: What Is the Difference Between Density-Dependent and Density-Independent Factors?
- Video: Threats to Sea Turtles

TEACHER NOTE **Connections: Crosscutting Concept: Scale, Proportion, and Quantity:** In this item, students understand that the significance of a limiting factor on a population can be dependent on the density of the population in which it occurs. Before they begin, discuss with students that patterns of relationships between biotic and abiotic factors that exist at some population densities may not exist at other densities. After students complete this item, deepen their understanding of this concept by asking them to categorize the factors listed according to whether they have no effect, a small effect, or a large effect on a dense population and on a dispersed population. Then, students can work with partners or in small groups to compare their answers, explaining the reasoning behind their categorizations.

TEI **Dependent or Independent?**

Lesson Question: What Are the Differences Between Exponential and Logistical Growth?

Recommended 30 minutes

- Core Interactive Text: What Are the Differences Between Exponential and Logistical Growth?
- Image: Doubling Time
- Image: Exponential Growth
- Video: Climate Change and Invasive Species
- Image: Logistic Growth
- Video: Reproductive Strategies

Lesson Question: How Is Population Growth Affected by Carrying Capacity?

Recommended 30 minutes

- Core Interactive Text: How Is Population Growth Affected by Carrying Capacity?
- Exploration: Describing Populations

TEACHER NOTE **Practice: Science and Engineering Practice: Developing and Using Models:** In Carrying Capacity, students develop and use a model to generate data to support an explanation of carrying capacity and predict the relationship between predator and prey populations in an ecosystem. After they complete this activity, instruct students to work in small groups to identify factors they could take into consideration to obtain a more accurate estimate of the ecosystem's carrying capacity. Each group should choose a factor and either revise the model to include it or write a plan for how the model could be revised.

- Hands-On Activity: Carrying Capacity

Lesson Question: How Are Age Structures Created?

`Recommended 30 minutes`

> **TEACHER NOTE Connections: Crosscutting Concept: Cause and Effect:** When examining age structure diagrams, students should understand that empirical evidence is required to make claims about the specific causes of population changes. To give students context for age structure diagrams and cause-and-effect relationships, organize them into small groups and assign each group a country. Students can use online resources to research the population of the country. For example, the United Nations Population Division has graphs showing population trends in all countries, including population pyramids. Instruct groups to describe their country's age structure diagram in 1950, 2015, and 2050. Then, instruct them to write a list of factors that could contribute to the changes they observe and how they could determine whether each of the potential factors they list could have caused the changes. If necessary, review the difference between cause and correlation with students.

- Core Interactive Text: How Are Age Structures Created?
- Image: Age Structure Diagram-Exponential
- Image: Age Structure Diagram-Declining
- Image: Age Structure Diagram-Stationary

Lesson Question: What Are the Differences Between Type I, Type II, and Type III Survivorship Curves?

`Recommended 30 minutes`

- Core Interactive What Are the Differences Between Type I, Type II, and Type III Survivorship Curves?
- Video: Predators and Prey

> **TEACHER NOTE Practice: Science and Engineering Practice: Constructing Explanations and Designing Solutions:** In these items, students apply scientific reasoning, theory, and mathematical models to link evidence to a claim about the relationship between survivorship curves and reproductive strategy. To extend this activity, have small groups of students discuss age structure, using evidence to support their claims about the relationships among age structure, survivorship curves, and reproductive strategies.

`TEI` The Curves of Survivorship
`TEI` What Is the Relationship?

Explore More Resources

Resources in Explore More Resources support differentiation within your classroom by

- providing additional visualization of content
- affording extension of content to those students ready for acceleration
- offering Lexile reading levels for reading passages

Online explorations and hands-on experiences are provided so that students can conduct virtual investigations, collect and design investigations, and collect and analyze data; these skills are essential to developing scientific understanding.

Explain `45–90 minutes`

In Explore, students
1. uncovered scientific understandings
2. conducted investigations
3. analyzed data, text, and other media resources
4. collected evidence to support their scientific explanation

In Explain, provide students with time to formally compose their scientific explanations around the Explain or student-generated questions using evidence collected from Explore.

Scientific explanations are student responses, either written or orally presented, that explain scientific phenomena based upon evidence. Developing a scientific explanation requires students to analyze and interpret data to construct meaning out of the data. There are three main components to the scientific explanation: the claim, the evidence, and the reasoning.

To help students to communicate their scientific explanations, allow them to utilize the multimedia creation tools such as Board Builder and Whiteboard. Remind them that they may upload image, audio, and video files using the "attach file" option to communicate their scientific explanations.

Students may construct their scientific explanations individually or within a small group of students. Students should communicate their explanations with other classmates, and provide constructive criticism and refine their explanations prior to submission to the teacher. If explanations are used as a formative assessment, you can provide additional feedback and comments to support students as they refine their explanations.

EXPLAIN

Explain the factors that impact squirrel populations.

Elaborate with STEM `45–135 minutes`

*Elaborate with STEM are optional extension resources available after students have demonstrated proficiency with standards addressed previously in the concept.

NGSS Components

SEP	CCC
■ Asking Questions and Defining Problems ■ Planning and Carrying Out Investigation ■ Analyzing and Interpreting Data ■ Using Mathematics and Computational Thinking ■ Constructing Explanations and Designing Solutions ■ Obtaining, Evaluating, and Communicating Information	■ Patterns ■ Cause and Effect ■ Systems and System Models ■ Stability and Change

STEM In Action `45 minutes`

STEM in Action ties the scientific concepts to real-world applications, with many connecting to STEM careers. Technology Enhanced Items (TEIs) expect students to critically read the Core Interactive Text (CIT) and review the provided media resources.

Applying Descriptions of Populations

- Core Interactive Text: Applying Descriptions of Populations
- Image: So Many Trees!
- Video: Does Sample Size Matter?
- Video: Tracking Bear Populations
- Video: Turtle Tracks
- Video: Population Dynamics

TEACHER NOTE This formative activity requires students to interpret data and relate it to the topics being learned. Students could work independently and then share and discuss their answers as a group.

- **TEI** Wolves in the Park
- **TEI** Wolves Outside the Park
- **TEI** Change in the Year 2005

STEM Project Starters

STEM Project Starters provide additional real-world contexts that require students to apply and extend their content knowledge related to the concept. STEM Project Starters can also serve as an alternative instructional hook presented at the beginning of the learning progression. The project can then be revisited throughout and at the end of the 5E learning cycle, for students to apply content knowledge.

STEM Project Starter: Carrying Capacity and Population Technology

`Recommended 90 minutes`

Why are transmitter tags a useful form of technology for monitoring?

> **TEACHER NOTE** This STEM project provides an opportunity for students to explore technology that biologists use for determining and monitoring a population's carrying capacity in an ecosystem. In order to enhance learning, you may wish to ensure that each group investigates a different form of technology or a different population. Have students work in pairs or groups of three to complete the activity. Afterward, have groups display their slideshows for the class and explain the technology they have researched.
>
> This project is a summative assessment.

- Activity: Engineering Design Sheet

STEM Project Starter Burmese Pythons Living in Florida?

`Recommended 90 minutes`

How did these snakes become invasive?

> **TEACHER NOTE** This summative assessment item requires students to read, research, interpret, and apply information they have learned to an example of an invasive species. This project is best completed by individuals or pairs. A class discussion can then occur to share ideas among students.

- Reading Passage: The Python Invasion

TEI Python Population

STEM Project Starter: How Do We Describe Populations?

`Recommended 30 minutes`

What are the factors involved in describing populations?

> **TEACHER NOTE** Students will need about 30 minutes to complete this summative assessment, where they will be required to interpret a graph and apply what they have learned to answer questions. This activity will work well with students working in pairs.

TEI Population Demographics
TEI Density
TEI Population Changes
TEI Survivorship Curves

Evaluate 45–90 minutes

Explain? Question:
Explain the factors that impact squirrel populations.

Lesson Questions (LQ):
- What are the factors of population demographics?
- What is the difference between density-dependent and density-independent factors?
- What are the differences between exponential and logistical growth?
- How is population growth affected by carrying capacity?
- How are age structures created?
- What are the differences between type I, type II, and type III survivorship curves?

Throughout instruction and the 5E learning cycle, you will have collected formative assessment data to drive the assignment of resources and experiences to students. Evaluate is intended to include summative assessment checks for proficiency. You can use the Explain and Lesson Questions for the concept as a summative assessment in a variety of ways such as these:

- Post each Lesson Question (LQ) in various locations in the classroom, and have small groups of students generate claim statements related to the Lesson Question (LQ). Other students can add to the claim, or refute the claim, during a gallery walk where they place additional pieces of evidence on each Lesson Question (LQ) poster.
- Assign small groups of students to each Lesson Question (LQ) and have the groups generate a poster, board, graphic, or piece of text that answers the question. Use a jigsaw approach and create a second set of groups that contain members from each Lesson Question (LQ) group to share their ideas.
- Ask students to return to their initial ideas for the Explain question and add additional details and evidence.

Encourage students to review the concept review and complete the Student Self-Check practice assessment prior to assigning the Summative Teacher Concept assessment.

- Student Review and Practice Assessment
- Teacher Concept Assessments

Biomes

The Five E Instructional Model
Science Techbook follows the 5E instructional model. This Model Lesson includes strategies for each of the 5Es. As you design the inquiry-based learning experience for students, be sure to collect data during instruction to drive your instructional decisions. Point-of-use teacher notes are also provided within each E-tab.

Engage   45–90 minutes

Engage Media Resources
The resources found in Engage are intended to stimulate students by exposing them to a phenomenon relevant to the content of the lesson. Engage also provides examples of relevant real-world applications that allow students to begin to make observations and relate the science content to their everyday lives. The Core Interactive Text (CIT) and media resources are carefully designed to prompt students to begin asking questions that they can investigate during the Explore phase of the lesson. They should also start collecting evidence to address the Explain question located at the bottom of the Engage page.

TEACHER NOTE **Investigative Phenomenon:** The investigative phenomenon for this concept is the grasslands that make up one of Earth's biomes. Students are introduced to the prairie ecosystem of North America and then to other grassland ecosystems around the world. Start by asking students about their pre-existing conceptions regarding prairies and the organisms that occupy them. Use the images provided in Explore More Resources to help stimulate their thinking. Once they have a firm idea of what the prairie ecosystem looks like, provide them with a world map and have them identify the location of the other grassland ecosystems that make up the grassland biome.

- Core Interactive Text: Thinking about Biomes
- Image: Buffalos: Animals of the Prairie
- Video: Fire and the Prairie
- Video: Life on the Prairie: Natural Grasses

Explain Question
The Explain question focuses students on gathering information in the Explore section. The Explain question can be used to

- Record what students already know related to the Explain question.
- Serve as a template or model for students to generate their own scientific questions.
- Collect evidence as students work through the lesson.
- Allow students to reflect on their growth before and after the lesson.

Explain
How does climate influence the characteristics of a biome?

- Image: Image: Tundra and Dwarf Trees

Engage Formative Assessment
Technology Enhanced Items (TEIs) found on the Engage page enable you to collect data on students' prior knowledge and identify the common misconceptions they may possess that are related to the topic of study. These items are designed as quick checks for understanding and allow each student one attempt at each question. You can use the data collected to decide whether to assign additional resources to the class, or determine what individual or groups of students may need reinforcement or accelerated learning, prior to completing the Explore portion of the lesson.

TEACHER NOTE Use this student response to evaluate students' prior knowledge of the concept. This activity can be conducted as a think-pair-share. Students will explain the difference between the four example biomes.

TEI Biome Comparison

Before You Begin
What Do I Already Know about Biomes?

TEACHER NOTE Use student responses to this formative assessment to determine whether students hold the misconception that biomes and ecosystems are the same. This activity can be conducted as a class discussion.

TEI Biomes and Ecosystems

TEACHER NOTE This is a formative assessment. This activity is intended to provide feedback on prior knowledge of geography and climate. Activity should be completed individually.

TEI World Climates

TEACHER NOTE Many students will have heard these biome names in middle school geography or science lessons or in contexts outside of school. This formative assessment is intended to provide feedback on students' misconceptions that biomes are only characterized by climate. Use this activity as class discussion to begin the lesson.

TEI Biomes Described

TEACHER NOTE Use this formative assessment to provide feedback on prerequisite knowledge of student understanding of the relationship between organisms within a food web. Students may have encountered this topic in middle school or in prior studies at the high school level. This activity can be used as a think-pair-share activity.

TEI **Effects on a Food Web**

- Video: Food Web
- Video: Ecological Pyramids
- Video: The Importance of Climate

Explore `135 minutes`

Lesson Questions (LQs):

1. What are the characteristics of Earth's terrestrial biomes?
2. What are the factors that determine Earth's aquatic biomes?
3. What are the characteristics of Earth's aquatic biomes?

Effective science instruction involves a student-centered rather than a teacher-centered approach. This can be accomplished either with Directed Inquiry or Guided Inquiry, depending on the needs and abilities of your class. Encourage students to select a variety of resources in their pursuit of answers as they work through Explore, with the end goal of constructing their scientific explanation in the Explain tab.

Directed Inquiry	Guided Inquiry
In Directed Inquiry, teachers provide students with a sequence of specified resources, challenging questions, and clear outcomes. Within this context students are given the opportunity to interact independently with each resource as prescribed by the teacher. Often different students groups can be guided through several different resources at the same time. For example, one group could work on a reading passage while a second group conducts a small-group Hands-On Activity with the teacher, and a third group is independently engaged with an online interactive resource.	In Guided Inquiry, students have independence to decide the scope and sequence of their investigations. Using resources from Tech book, students determine for themselves which resources they will Explore to answer the Lesson Questions. It is important to note that each student will choose multiple resources, but no one student is expected to use all the resources available. Students also determine the order in which to explore these resources and how to record their findings.

NGSS Components

SEP	CCC
■ Connections to Nature of Science ■ Engaging in Argument from Evidence	■ Cause and Effect ■ Stability and Change ■ Patterns ■ Structure and Function

Lesson Question: What Are the Characteristics of Earth's Terrestrial Biomes?

Recommended 45 minutes

TEACHER NOTE Connections: Crosscutting Concept: Cause and Effect: In this concept, students investigate the characteristics of Earth's terrestrial and aquatic biomes. As they learn about each biome, they suggest cause and effect relationships to explain and predict behaviors in complex natural systems, such as how the climate and soil of a particular terrestrial biome have caused plants and animals to develop specific adaptations to survive there.

Use a strategy such as "25 Things You Didn't Know" to extend and deepen students' knowledge of Earth's biomes. For information on this strategy, go to the Professional Learning tab of Techbook and click on Strategies & Resources. Then click on Spotlight on Strategies (SOS). "25 Things You Didn't Know" is found under "Research."

As students read and comprehend complex texts, view the videos, and complete the interactives, labs, and other Hands-On Activities, have them summarize and obtain scientific and technical information. Students will use this evidence to support their initial ideas on how to answer the Explain Question or their own question they generated during Engage. Have students record their evidence using "My Notebook.

- Core Interactive Text: Defining a biome

TEACHER NOTE Students may think biomes are synonymous with ecosystems. There are slight differences between the two terms. For the most part, ecosystems are smaller than biomes. Also, a biome may be scattered around Earth and made up of several geographically separate ecosystems.

- Image: Terrestrial Biomes

TEACHER NOTE Misconception: Students may assume that all biomes are characterized just by the climate of the area. However, all biomes are characterized by both abiotic and biotic factors.

- Image: Earth's Terrestrial Biomes
- Exploration: Terrestrial Biomes

TEACHER NOTE Science and Engineering Practices: Connections to Nature of Science: Scientific Knowledge is Open to Revision in Light of New Evidence: Have students review the summary of terrestrial biomes provided in the table. Emphasize the importance of climate in determining the flora and fauna within a biome. Next have seven small groups each research one specific biome using Techbook and other library and online resources Have each group create a Board or poster detailing examples of adaptations the animals and plants have, to meet the climatic conditions in the biome. Have students present or otherwise share their findings.

- Video: The Sahara Desert
- Video: Tundra
- Video: Taiga or Boreal Forest
- Video: Tropical Rain Forest
- Video: Temperate Deciduous Forest
- Video: Temperate Grasslands

Formative Assessment:

Throughout Explore, Technology Enhanced Items (TEIs) are embedded as multi-dimensional formative checks for understanding. You can use the data they provide to

- assign additional support
- extend learning
- design additional learning tasks to clarify student misconceptions

The Explore TEIs provide students with three attempts to demonstrate their proficiency. Scaffolded feedback is provided for each attempt. If a student does not achieve proficiency by the third attempt, a media asset is provided as an additional learning opportunity.

TEACHER NOTE Connections: Crosscutting Concept: Stability and Change: Each terrestrial biome is a unique system with plant and animal life that has specific adaptations for survival in that system. In the following item, students analyze the biome systems to define their boundaries and the impact of these on the survival of organisms. Extend these items by having a class discussion about how animal and plant adaptations relate to the soil and climate of each biome.

TEI Are Tundras Deserts?

TEI What is this Biome

Lesson Question: What are the Factors that Determine Earth's Aquatic Biomes?

Recommended 45 minutes

TEACHER NOTE Science and Engineering Practice: Engaging in Argument from Evidence: In this lesson question students read and evaluate multiple sources of information as they learn about the different aquatic biomes and the factors that make these biomes distinct from one another. Have students construct a table in which they compare aspects of marine aquatic biomes and freshwater aquatic biomes. Students can then draw upon what they have learned to identify unique characteristics and the types of organisms that inhabit these biomes. Finally, students should communicate their findings and engage in discussions regarding the impact of human activity on these biomes and the implications of these effects.

- Core Interactive Text: What Abiotic Factors Affect Life in Aquatic Biomes?

TEACHER NOTE Misconception: Students may think that the photic zone always extends to the same depth. The photic zone varies with conditions, including the time of day, angle of the sun to the ocean, and turbidity of the water.

TEACHER NOTE Misconception: Students may think that planktonic organisms are microscopic. Most plankton species are microscopic, but the term *planktonic* refers to organisms that drift. There are many macroscopic planktonic organisms.

- Image: Ocean Food Web
- Image: Phytoplankton
- Video: Phytoplankton in the Ocean
- Image: Melting Sea Ice
- Animation: Trophic Levels

TEACHER NOTE Connections Crosscutting Concept: Patterns: This item requires students to classify bodies of water as either marine or freshwater. Once students have completed this task, lead a discussion about scale and salt concentrations, which differentiate these two types of aquatic biomes. Discuss the potential overlap among various aquatic biomes to create a segue for moving into the next section of the text, which deals with abiotic factors.

TEI Marine or Freshwater

Lesson Question: What are the characteristics of Earth's aquatic biomes?

Recommended 45 minutes

- Core Interactive Text: Fresh Water Biomes
- Image: Glaciers Provide Fresh Water
- Image: Flood Plain
- Video: Bull Sharks and Crocodiles
- Video: The Amazon River Basin

TEACHER NOTE Connections Crosscutting Concept: Patterns: In this item, students will recognize that observed patterns in nature, such as those that differentiate the zones of streams and rivers, help guide how we classify and organize the natural world and can prompt questions about the relationships between the zones. Once students have completed this item, lead a discussion about the differences in life that can be sustained in the various zones. This would also be an excellent opportunity to discuss trade-offs between the oxygen and nutrient levels in the source and floodplain zones.

TEI Nutrient Densities and Oxygen Levels

TEACHER NOTE To make the differences in nutrient levels in lakes more relatable to students, find photos of lakes that fall into the three categories: oligotrophic, eutrophic, and mesotrophic. Provide images and discuss the types of life found in each of these lake types to help make the differences clear.

- Core Interactive Text: Lakes and Ponds
- Image: Lakes and Ponds
- Image: Cyprus Swamps
- Video: Okavango Kavango Delta
- Core Interactive Text: Marine Biomes
- Image: Mangroves
- Image: Sea Lions on Rocky Coast
- Video: Adaptations of Reef Fish
- Video: The Ocean Food Web
- Video: Ocean Zones
- Video: Life on the Sea Floor
- Video: Estuaries

TEACHER NOTE Connections: Crosscutting Concept: Structure and Function: This item requires that students establish the structural, abiotic factors associated with various marine biomes. The functions and properties of marine biomes can be inferred by learning about their overall structure, and the ways their various components are shaped and used.

TEI Marine Biomes

Explore More Resources

Resources in Explore More Resources support differentiation within your classroom by

- providing additional visualization of content
- affording extension of content to those students ready for acceleration
- offering Lexile reading levels for reading passages

Online explorations and hands-on experiences are provided so that students can conduct virtual investigations, collect and design investigations, and collect and analyze data; these skills are essential to developing scientific understanding.

Explain `45–90 minutes`

In Explore, students
1. uncovered scientific understandings
2. conducted investigations
3. analyzed data, text, and other media resources
4. collected evidence to support their scientific explanation

In Explain, provide students with time to formally compose their scientific explanations around the Explain or student-generated questions using evidence collected from Explore.

Scientific explanations are student responses, either written or orally presented, that explain scientific phenomena based upon evidence. Developing a scientific explanation requires students to analyze and interpret data to construct meaning out of the data. There are three main components to the scientific explanation: the claim, the evidence, and the reasoning.

To help students to communicate their scientific explanations, allow them to utilize the multimedia creation tools such as Board Builder and Whiteboard. Remind them that they may upload image, audio, and video files using the "attach file" option to communicate their scientific explanations.

Students may construct their scientific explanations individually or within a small group of students. Students should communicate their explanations with other classmates, and provide constructive criticism and refine their explanations prior to submission to the teacher. If explanations are used as a formative assessment, you can provide additional feedback and comments to support students as they refine their explanations.

EXPLAIN

How does climate influence the characteristics of a biome?

Elaborate with STEM `45–180 minutes`

*Elaborate with STEM are optional extension resources available after students have demonstrated proficiency with standards addressed previously in the concept.

NGSS Components

SEP	CCC
■ Asking Questions and Defining Problems ■ Developing and Using Models ■ Using Mathematics and Computational Thinking ■ Constructing Explanations and Designing Solutions ■ Engaging in Argument from Evidence	■ Cause and Effect ■ Energy and Matter ■ Patterns ■ Stability and Change ■ Systems and System Models

STEM In Action `45 minutes`

STEM in Action ties the scientific concepts to real-world applications, with many connecting to STEM careers. Technology Enhanced Items (TEIs) expect students to critically read the Core Interactive Text (CIT) and review the provided media resources.

Applying Biomes

- Core Interactive Text: Applying Biomes
- Image: Coral Bleaching
- Video: Bleaching at Rainbow Reef
- Video: To the Deepest Part of the Ocean
- Image: Robotic Submersible
- Video: The First Test Dive
- Video: Threats to Coral Reef

TEACHER NOTE This formative assessment is designed to show the relationship between increasing water temperature and coral bleaching events. Present this material to students toward the end of the discussion on coral bleaching. Have students work individually to answer the questions.

For the Bleaching Event Update item, students are asked to locate more recent data to determine whether the correlation between sea surface temperature and bleaching events still holds true. Coral Reef Watch, part of NOAA's National Environmental and Satellite and Information Service (NESDIS), regularly publishes on the Internet data about coral bleaching events. The Great Barrier Reef Marine Park Authority, part of the Australian government, also posts information about bleaching events.

- **TEI** Frequency of Bleaching
- **TEI** Possible Causes
- **TEI** Bleaching Event Update

STEM Project Starters

STEM Project Starters provide additional real-world contexts that require students to apply and extend their content knowledge related to the concept. STEM Project Starters can also serve as an alternative instructional hook presented at the beginning of the learning progression. The project can then be revisited throughout and at the end of the 5E learning cycle, for students to apply content knowledge.

STEM Project Starter: Climate Change by the Numbers
Recommended 45 minutes

What is causing the melting of polar ice caps and the destruction of the habitat that is home to many Arctic animal species?

> **TEACHER NOTE** This summative assessment should be completed individually so that all students get practice constructing dual-axis graphs. Student may need conversion factors to change precipitation and temperature into metric. Assign students a location so that every biome is represented by the class. Once students have all completed their graphs, discuss the biome that each graph represents.

- **TEI** Collect and Show Data
- **TEI** Interpret Climate Data

STEM Project Starter: Biome Challenge
Recommended 45 minutes

What resources are available in biomes in which humans do not traditionally live?

> **TEACHER NOTE** This activity could be completed individually or in pairs. The purpose of this project is a summative assessment to assess student understanding of the concept. The teacher may choose to assign the locations so that a variety is represented in the class.

- **TEI** Biome Challenge

STEM Project Starter: Project: Biotic versus Abiotic
Recommended 135 minutes

What are some of the factors that can alter the composition of and organisms within aquatic biomes?

> **TEACHER NOTE** This project may be used as a formative assessment and as a follow-on to the two Hands-On Activity: Abiotic and Biotic Factors.
>
> This project is best completed in small groups. Consider your students and types of data collected during the labs when deciding how to organize student groups. It may make sense to organize students into the same groups that collected the data during the field experience. Or it may make more sense to recombine students so that each group has access to multiple data sets for comparison and analysis.
>
> Students' results depend on the data they collected during the labs. Their thought processes and ability to ask questions and propose relationships based on data are more important indicators or success than any clearly observable correlation between factors from their data.

- Hands-On Activity: Abiotic Factors
- Hands-On Activity: Biotic Factors

STEM Project Starter: Project: The Microbiome of the Human Body

Recommended 90 minutes

How can microbes living on and in the human body be considered to be a biome?

TEACHER NOTE This summative assessment provides students with the opportunity to consider the scale of ecosystems. In particular, students should consider the human body as an ecosystem with the same inputs, outputs, and interactions as a woodland pond or seashore. For example, the video, Bacteria: Bad and Good shows that interactions between good and bad bacteria are akin to predator-prey interactions in these more familiar ecosystems. The aim is for students to compare and contrast Earth's spheres with the body's structure and function, with specific focus on the similarities and differences between the human body and its microbiome, and Earth's ecosystems.

- Video: Bacteria: Bad and Good

TEACHER NOTE In these items, students are asked to conduct research before attempting the questions. They will research online resources and summarize their findings. The aim is for students to cite evidence explaining why the microbiome helps to determine the health of the body. They are not asked to identify specific microbes, but they could do so if they want. If time allows, groups can share their research to create a map of the human body with a taxonomy of microorganisms that comprise the human microbiome.

- **TEI** Research a Microbe
- **TEI** The Body as an Ecosystem
- **TEI** Compare and Contrast

Evaluate `45–90 minutes`

Explain Question:
How does climate influence the characteristics of a biome?

Lesson Questions (LQ):
- What are the characteristics of Earth's terrestrial biomes?
- What are the factors that determine Earth's aquatic biomes?
- What are the characteristics of Earth's aquatic biomes?

Throughout instruction and the 5E learning cycle, you will have collected formative assessment data to drive the assignment of resources and experiences to students. Evaluate is intended to include summative assessment checks for proficiency. You can use the Explain and Lesson Questions for the concept as a summative assessment in a variety of ways such as these:

- Post each Lesson Question (LQ) in various locations in the classroom, and have small groups of students generate claim statements related to the Lesson Question (LQ). Other students can add to the claim, or refute the claim, during a gallery walk where they place additional pieces of evidence on each Lesson Question (LQ) poster.
- Assign small groups of students to each Lesson Question (LQ) and have the groups generate a poster, board, graphic, or piece of text that answers the question. Use a jigsaw approach and create a second set of groups that contain members from each Lesson Question (LQ) group to share their ideas.
- Ask students to return to their initial ideas for the Explain question and add additional details and evidence.

Encourage students to review the concept review and complete the Student Self-Check practice assessment prior to assigning the Summative Teacher Concept assessment.

- Student Review and Practice Assessment
- Teacher Concept Assessments

Energy for Life

The Five E Instructional Model

Science Techbook follows the 5E instructional model. This Model Lesson includes strategies for each of the 5Es. As you design the inquiry-based learning experience for students, be sure to collect data during instruction to drive your instructional decisions. Point-of-use teacher notes are also provided within each E-tab.

Engage   45–90 minutes

Engage Media Resources

The resources found in Engage are intended to stimulate students by exposing them to a phenomenon relevant to the content of the lesson. Engage also provides examples of relevant real-world applications that allow students to begin to make observations and relate the science content to their everyday lives. The Core Interactive Text (CIT) and media resources are carefully designed to prompt students to begin asking questions that they can investigate during the Explore phase of the lesson. They should also start collecting evidence to address the Explain question located at the bottom of the Engage page.

TEACHER NOTE Investigative Phenomenon: Students are introduced to the role of energy in living organisms through the example of running a marathon. Start by asking students about their own experiences of running long-distance races and training. What do they need to do before a race? What types of food would they eat and why? If they use the term "energy," ask them what they think energy is. Is there only one form of energy? If not, what forms of energy would they associate with participating in a race? Students will most likely suggest that the runners use food to provide themselves with energy, and that movement is a type of energy since heat (thermal) energy is generated, which makes them sweat. Encourage them to explain how food contains energy, where that energy comes from, and how it is converted to other forms of energy. Use their ideas, and the pre-assessment items in Engage, to gauge the level of students' understanding of the nature of energy and how biological systems utilize it.

- Core Interactive Text: Thinking about the Energy for Life
- Image: Running a Marathon
- Video: Food Energy
- Video: Energy for Life

Explain Question

The Explain question focuses students on gathering information in the Explore section. The Explain question can be used to

- Record what students already know related to the Explain question.
- Serve as a template or model for students to generate their own scientific questions.
- Collect evidence as students work through the lesson.
- Allow students to reflect on their growth before and after the lesson.

Explain

Explain how and why energy is important to living things.

> ■ Image: Ocean Food Web

Engage Formative Assessment

Technology Enhanced Items (TEIs) found on the Engage page enable you to collect data on students' prior knowledge and identify the common misconceptions they may possess that are related to the topic of study. These items are designed as quick checks for understanding and allow each student one attempt at each question. You can use the data collected to decide whether to assign additional resources to the class, or determine what individual or groups of students may need reinforcement or accelerated learning, prior to completing the Explore portion of the lesson.

> **TEACHER NOTE** Use this student response to evaluate students' prior knowledge of the concept. The Model Lesson provides information on common student misconceptions. Have students answer the question and then revisit it later in the lesson.

TEI Energy Source

Before You Begin

What Do I Already Know about the Energy for Life?

> **TEACHER NOTE** This formative assessment item is intended to provide the teacher with feedback on prior knowledge of this topic and to identify existing misconceptions. In middle school, students should have learned about the various organic molecules and should be familiar with the more common ones. Use this as a lesson opener before a discussion of the different organic molecules. This activity will allow students to review what they already know about organic molecules and to set up their learning for how carbohydrates are used for energy.

TEI The Energy of Carbohydrates

> **TEACHER NOTE** This is a formative assessment item that assesses students' knowledge of dimensional analysis. Use this as a lesson opener before reviewing metric conversions. Have students work on it together and then review as a class. Students selecting A are confused about how many millimeters there are in a meter and moved the decimal point the wrong way. Students selecting B are confused about how many millimeters there are in a meter and did not move the decimal point the correct way or enough places. Students selecting C have a misconception about how many millimeters there are in a meter and did not move the decimal point enough places to the right.

TEI Math and Energy Conversions

TEACHER NOTE This formative assessment item is intended to provide the teacher with feedback on prior knowledge of this topic. Use this as a lesson opener to review the topics. Have students work together to determine the answers and then review as a class. Its main focus is to determine whether students remember the main parts of cellular respiration and photosynthesis. Review of this information will be useful as they learn more about the energy both of these processes produce for living things.

TEI Energy Processes

TEACHER NOTE Use this formative assessment item as a quick review in between discussion points at the beginning of the lesson. Have students raise their hands to answer. Students should be aware of the relationship between energy and matter and that energy is needed for chemical reactions to take place.

TEI Energy and Reactions

- Video: Food Chains, Food Webs, and Trophic Levels
- Video: Metric Conversions
- Video: Photosynthesis and Cellular Respiration

Explore `180 minutes`

Lesson Questions (LQs):

1. What is energy and how does it contribute to maintaining life on Earth?
2. What are kinetic and potential energy?
3. What are the laws of thermodynamics?
4. What is the difference between endergonic and exergonic reactions?
5. What is the difference between oxidation and reduction reactions?
6. What Is the Role of Photosynthesis and Respiration in the Carbon Cycle?

Effective science instruction involves a student-centered rather than a teacher-centered approach. This can be accomplished either with Directed Inquiry or Guided Inquiry, depending on the needs and abilities of your class. Encourage students to select a variety of resources in their pursuit of answers as they work through Explore, with the end goal of constructing their scientific explanation in the Explain tab.

Directed Inquiry	Guided Inquiry
In Directed Inquiry, teachers provide students with a sequence of specified resources, challenging questions, and clear outcomes. Within this context students are given the opportunity to interact independently with each resource as prescribed by the teacher. Often different students groups can be guided through several different resources at the same time. For example, one group could work on a reading passage while a second group conducts a small-group Hands-On Activity with the teacher, and a third group is independently engaged with an online interactive resource.	In Guided Inquiry, students have independence to decide the scope and sequence of their investigations. Using resources from Techbook, students determine for themselves which resources they will Explore to answer the Lesson Questions. It is important to note that each student will choose multiple resources, but no one student is expected to use all the resources available. Students also determine the order in which to explore these resources and how to record their findings.

NGSS Components

SEP	CCC
■ Developing and Using Models ■ Engaging in Argument from Evidence	■ Energy and Matter

Lesson Question: What is Energy and How Does It Contribute to Maintaining Life on Earth?

Recommended 70 minutes

TEACHER NOTE **Crosscutting Concepts: Energy and Matter:** In this concept, students will understand the role that energy plays in living systems. They will learn about energy changes in a system. They also will learn that energy cannot be created nor destroyed. As students learn about energy, have them summarize the information using the Journals strategy. The Journals strategy is found on the Professional Learning tab. Click on Strategies & Resources, then click on Spotlight on Strategies (SOS). Now click on Vocabulary Development, then click on Spotlight on Strategies: Journals.

- Core Interactive Text: What is Energy and How Does It Contribute to Maintaining Life on Earth?
- Video: Introduction to Matter and Energy
- Video: The Limits of Life
- Video: An Introduction to Energy and Work
- Video: Energy as Work
- Hands-On Lab: Energy in Food

Formative Assessment:

Throughout Explore, Technology Enhanced Items (TEIs) are embedded as multi-dimensional formative checks for understanding. You can use the data they provide to

- assign additional support
- extend learning
- design additional learning tasks to clarify student misconceptions

The Explore TEIs provide students with three attempts to demonstrate their proficiency. Scaffolded feedback is provided for each attempt. If a student does not achieve proficiency by the third attempt, a media asset is provided as an additional learning opportunity.

TEACHER NOTE **Crosscutting Concepts: Energy and Matter: Flows, Cycles, and Conservation:** In this item, students will analyze different life processes to match them with descriptions of the way energy flows into, through, and within that system. The energy uses of each process may not be obvious to students at first. Review other concepts with them that they have learned in Biology, including DNA replication, digestion, respiration, and the nervous system to help them recall how energy is used in each of these processes. Have small groups write tweets to summarize how each system uses energy using the strategy Tweet, Tweet! Access this strategy on Techbook by clicking the Professional Learning tab. Click on Strategies & Resources, then Spotlight on Strategies (SOS). Tweet, Tweet! is found underneath Summarizing.

TEI Energy in Life

Lesson Question: What Are Kinetic and Potential Energy?   `Recommended 35 minutes`

- Core Interactive Text: What Are Kinetic and Potential Energy?
- Video: Kinetic Energy

TEACHER NOTE Science and Engineering Practice: Developing and Using Models: In this item, students will use a rollercoaster to model the changes in kinetic and potential energy of an object moving in two dimensions. They will use a model based on evidence to illustrate and predict the relationships between systems. Help students understand the movement of the object by drawing the roller coaster on the board. Have them move a ball along the rollercoaster to model the movement of the rollercoaster car. At each step, ask students how the kinetic and potential energies of the car are changing. After students complete this item, challenge them to apply the information in the graph to biological processes by having them write a paragraph that explains potential and kinetic energy in living things.

TEI Rollercoaster Energy

TEACHER NOTE Misconception: Students may believe that living organisms are not governed by the laws of thermodynamics. The laws of thermodynamics apply to everything in the Universe.

Lesson Question: What Are the Laws of Thermodynamics?   `Recommended 20 minutes`

- Core Interactive Text: What Are the Laws of Thermodynamics?
- Video: The Three Laws of Thermodynamics
- Video: Transfer of Energy Between Bouncing Balls

Lesson Question: What Is the Difference between Endergonic and Exergonic Reactions?

Recommended 20 minutes

- Core Interactive Text: What Is the Difference between Endergonic and Exergonic Reactions?
- Image: ATP

TEACHER NOTE Emphasize to students that endergonic and exergonic reactions are not the same as endothermic and exothermic reactions. Although many exothermic reactions are also exergonic, such as burning wood, many are not, such as dissolving table salt in water. Although dissolving salt in water is spontaneous, it is endothermic. Have students demonstrate their understanding of these reactions using the Snowball Fight strategy. Students should write an example of a reaction on a scrap of paper, then crumple that paper and toss it in the air. Students then grab a new scrap of paper, uncrumple it, and try to correctly name the type of reaction described in the example. Access this strategy by clicking on the Professional Learning tab on Science Techbook. Click on Strategies & Resources, then Spotlight on Strategies (SOS). Snowball Fight is found underneath Key Ideas and Details.

- Video: ATP: The Energy Currency

Lesson Question: What Is the Difference between Oxidation and Reduction Reactions?

Recommended 35 minutes

- Core Interactive Text: What Is the Difference between Oxidation and Reduction Reactions
- Video: Redox Reactions

TEACHER NOTE Science and Engineering Practice: Engaging in Argument from Evidence: In this item, students explain why the energy of an electron decreases as photosynthesis uses the energy in the electron to reduce carbons. They construct, use, and present a written argument based on data and evidence. Use the Twenty Questions strategy to help them understand the information presented in the graph. As they examine the graph, have them pose questions about the graph. The Twenty Questions strategy is found on the Professional Learning tab. Click on Strategies & Resources, then click on Spotlight on Strategies (SOS). Now click on Inference and Prediction, then click on Spotlight on Strategies: Twenty Questions.

TEI Stopped Reaction

Lesson Question: What Is the Role of Photosynthesis and
Respiration in the Carbon Cycle?

Recommended 40 minutes

- Core Interactive Text: What Is the Role of Photosynthesis and Respiration in the Carbon Cycle?
- Video: The Carbon Cycle
- Video: Fast Carbon Cycle
- Image: Photosynthesis
- Video: The Process of Cellular Respiration
- Video: Photosynthesis and Cell Respiration

TEACHER NOTE This item assesses students' ability to analyze the cause-and-effect relationships in the carbon cycle. Students will use what they have learned about the smaller scale processes within the carbon cycle to predict how changes in the quantities of inputs and the number of organisms would affect the overall quantity of carbon in the atmosphere. To provide extra support, have students write the overall reactions for photosynthesis and respiration. Then, ask leading questions to help students determine the effect of changing quantities of reagents and populations of organisms. For example, Would increasing the population organisms increase or decrease the rate of respiration? Would increasing the rate of respiration increase the quantity of reagents or products in the reaction? How would this affect the amount of carbon in the atmosphere?

TEI Fast Carbon Cycle

Explore More Resources

Resources in Explore More Resources support differentiation within your classroom by

- providing additional visualization of content
- affording extension of content to those students ready for acceleration
- offering Lexile reading levels for reading passages

Online explorations and hands-on experiences are provided so that students can conduct virtual investigations, collect and design investigations, and collect and analyze data; these skills are essential to developing scientific understanding.

Explain `45–90 minutes`

In Explore, students
1. uncovered scientific understandings
2. conducted investigations
3. analyzed data, text, and other media resources
4. collected evidence to support their scientific explanation

In Explain, provide students with time to formally compose their scientific explanations around the CYE or student-generated questions using evidence collected from Explore.

Scientific explanations are student responses, either written or orally presented, that explain scientific phenomena based upon evidence. Developing a scientific explanation requires students to analyze and interpret data to construct meaning out of the data. There are three main components to the scientific explanation: the claim, the evidence, and the reasoning.

To help students to communicate their scientific explanations, allow them to utilize the multimedia creation tools such as Board Builder and Whiteboard. Remind them that they may upload image, audio, and video files using the "attach file" option to communicate their scientific explanations.

Students may construct their scientific explanations individually or within a small group of students. Students should communicate their explanations with other classmates, and provide constructive criticism and refine their explanations prior to submission to the teacher. If explanations are used as a formative assessment, you can provide additional feedback and comments to support students as they refine their explanations.

CAN YOU EXPLAIN?
Explain how and why energy is important to living things.

Elaborate with STEM `45–135 minutes`

*Elaborate with STEM are optional extension resources available after students have demonstrated proficiency with standards addressed previously in the concept.

NGSS Components

SEP	CCC
■ Obtaining, Evaluating, and Communicating Information ■ Asking Questions and Defining Problems ■ Planning and Carrying Out Investigation ■ Using Mathematics and Computational Thinking ■ Constructing Explanations and Designing Solutions	■ Cause and Effect ■ Structure and Function

STEM In Action `45 minutes`

STEM in Action ties the scientific concepts to real-world applications, with many connecting to STEM careers. Technology Enhanced Items (TEIs) expect students to critically read the Core Interactive Text (CIT) and review the provided media resources.

Applying Energy for Life

- Core Interactive Text: Applying Energy for Life
- Image: Calorie Label
- Video: BMI
- Video: Biomass Energy
- Video: Making Gasoline from Sawdust
- Video: Oil from Algae
- Video: Corn Grown for Biofuel

TEACHER NOTE Use this summative assessment as a concluding activity after discussing the concepts of BMI, body weight, and energy. Have the students work on it together and then share their findings with the class.

TEI Burn Calories

TEI Negative to Positive?

TEI Humans

STEM Project Starters

STEM Project Starters provide additional real-world contexts that require students to apply and extend their content knowledge related to the concept. STEM Project Starters can also serve as an alternative instructional hook presented at the beginning of the learning progression. The project can then be revisited throughout and at the end of the 5E learning cycle, for students to apply content knowledge.

STEM Project Starter: The Mighty Mitochondria

Recommended 60 minutes

How does the anatomy of a mitochondria support its function?

> **TEACHER NOTE** This STEM project relates to cellular respiration and energy. As students build their models, they will gain a better understanding of where energy is produced within the mitochondrion. Their detailed labels will serve as a review of the organelle. This project should be done alone and as a take-home activity. Provide students with adequate time to complete their models.

TEI Explaining My Model

STEM Project Starter: Get Moving!

Recommended 90 minutes

How does potential energy convert to kinetic energy?

> **TEACHER NOTE** This is a formative assessment activity. Students should submit their reports at least twice—once when they have designed their procedure and once when they have conducted the investigation. This activity will give students hands-on experience with an energy conversion. They will then need to dig into it further to see how changing potential to kinetic energy is useful in nature. Have students perform this activity in pairs or small groups, as materials may be limited.

TEI Get Moving

STEM Project Starter: Calculating BMI

Recommended 45 minutes

What does BMI tell you about a person?

> **TEACHER NOTE** This project will give students exposure to BMI and how to calculate it. They will use the graph as a tool but should also do some additional research to check their accuracy. Have students work in pairs or small groups to complete the activity, so they can validate each other's answers.

TEI Calculating BMI

Evaluate `45–90 minutes`

Explain Question:
Explain how and why energy is important to living things.

Lesson Questions (LQ):
- What is energy and how does it contribute to maintaining life on Earth?
- What are kinetic and potential energy?
- What are the laws of thermodynamics?
- What is the difference between endergonic and exergonic reactions?
- What is the difference between oxidation and reduction reactions?
- What Is the Role of Photosynthesis and Respiration in the Carbon Cycle?

Throughout instruction and the 5E learning cycle, you will have collected formative assessment data to drive the assignment of resources and experiences to students. Evaluate is intended to include summative assessment checks for proficiency. You can use the CYE and Lesson Questions for the concept as a summative assessment in a variety of ways such as these:

- Post each Lesson Question (LQ) in various locations in the classroom, and have small groups of students generate claim statements related to the Lesson Question (LQ). Other students can add to the claim, or refute the claim, during a gallery walk where they place additional pieces of evidence on each Lesson Question (LQ) poster.
- Assign small groups of students to each Lesson Question (LQ) and have the groups generate a poster, board, graphic, or piece of text that answers the question. Use a jigsaw approach and create a second set of groups that contain members from each Lesson Question (LQ) group to share their ideas.
- Ask students to return to their initial ideas for the Explain question and add additional details and evidence.

Encourage students to review the concept review and complete the Student Self-Check practice assessment prior to assigning the Summative Teacher Concept assessment.

- Student Review and Practice Assessment
- Teacher Concept Assessments

Photosynthesis

The Five E Instructional Model

Science Techbook follows the 5E instructional model. This Model Lesson includes strategies for each of the 5Es. As you design the inquiry-based learning experience for students, be sure to collect data during instruction to drive your instructional decisions. Point-of-use teacher notes are also provided within each E-tab.

Engage   45–90 minutes

Engage Media Resources

The resources found in Engage are intended to stimulate students by exposing them to a phenomenon relevant to the content of the lesson. Engage also provides examples of relevant real-world applications that allow students to begin to make observations and relate the science content to their everyday lives. The Core Interactive Text (CIT) and media resources are carefully designed to prompt students to begin asking questions that they can investigate during the Explore phase of the lesson. They should also start collecting evidence to address the Explain question located at the bottom of the Engage page.

> **TEACHER NOTE** **Investigative Phenomenon:** In Engage, students are presented with information about the sun's transfer of energy to Earth and how this provides the energy for many of Earth's systems and living organisms. After showing the short video clip, Sun, have students discuss in small groups: How could the sun, itself so inhospitable to life, possibly be essential to life's very existence? Ask them to map out how they think energy generated by the sun's nuclear reactions could provide them with lunch, and then use the remainder of Engage to help them consolidate and assess their ideas about the light-life connection before delving into the details of the process of photosynthesis.

- Core Interactive Text: A Light Lunch
- Video: Sun
- Video: What Do Plants Need?
- Video: The Sun as an Energy Source
- Video: Photosynthesis and Introduction

> **TEACHER NOTE** Students should answer this using the assessment item and revisit their answers when they start the Explain part of the concept.

Explain Question

The Explain question focuses students on gathering information in the Explore section. The Explain question can be used to

- Record what students already know related to the Explain question.
- Serve as a template or model for students to generate their own scientific questions.
- Collect evidence as students work through the lesson.
- Allow students to reflect on their growth before and after the lesson.

Explain

Can you explain the steps involved in converting sunlight into the energy that is stored in food and compare the processes of cellular respiration and photosynthesis?

- Image: From Sunlight to Food

TEACHER NOTE Use this student response to evaluate students' prior knowledge of the concept. The Model Lesson provides information on common student misconceptions. This may be used first as a small-group discussion to allow students to recall as much as they can about the topic. Individuals should enter their response to give you the best idea of each student's prior knowledge.

TEI Your Ideas

Engage Formative Assessment

Technology Enhanced Items (TEIs) found on the Engage page enable you to collect data on students' prior knowledge and identify the common misconceptions they may possess that are related to the topic of study. These items are designed as quick checks for understanding and allow each student one attempt at each question. You can use the data collected to decide whether to assign additional resources to the class, or determine what individual or groups of students may need reinforcement or accelerated learning, prior to completing the Explore portion of the lesson.

Before You Begin

What Do I Already Know about Photosynthesis?

TEACHER NOTE This activity is intended to provide the teacher with feedback on prior knowledge of this topic. In middle school, students should have learned that photosynthesis has reactants and products and should have an understanding of what these are and how they are related. Use this as a think-pair-share activity.

This is a formative assessment.

TEI Reactants and Products

TEACHER NOTE This activity will help identify students who may not yet understand the concept of photosynthesis and the role of chlorophyll and the sun.

Students choosing A may not understand how we see color and the different waves of visible light. Students selecting C or D may not understand that it is the visible light waves that are needed for photosynthesis. Students selecting E may not understand that the energy must be absorbed and stored in the plant. Use this as a think-pair-share activity.

This is a formative assessment.

TEI Sunlight and Photosynthesis

TEACHER NOTE This activity is intended to provide the teacher with feedback on prior knowledge of this topic and possible misconceptions students may have about chlorophyll and the color it can appear. Students may work with partners to discuss the possible definitions and the matching terms. Students will have to separate correct definitions from those that do not apply to the terms listed. The Model Lesson has remediation available for students struggling to master the concepts.

This is a formative assessment.

TEI Terms of Photosynthesis

- Video: Inside Plants: Leaves Overview and Introduction
- Video: Science and the City: Energy
- Video: Science in Progress: Plant Cells

Explore `90 minutes`

Lesson Questions (LQs):

1. How are photosynthesis and cellular respiration related?
2. How does the process of photosynthesis work?

Effective science instruction involves a student-centered rather than a teacher-centered approach. This can be accomplished either with Directed Inquiry or Guided Inquiry, depending on the needs and abilities of your class. Encourage students to select a variety of resources in their pursuit of answers as they work through Explore, with the end goal of constructing their scientific explanation in the Explain tab.

Directed Inquiry	Guided Inquiry
In Directed Inquiry, teachers provide students with a sequence of specified resources, challenging questions, and clear outcomes. Within this context students are given the opportunity to interact independently with each resource as prescribed by the teacher. Often different students groups can be guided through several different resources at the same time. For example, one group could work on a reading passage while a second group conducts a small-group Hands-On Activity with the teacher, and a third group is independently engaged with an online interactive resource.	In Guided Inquiry, students have independence to decide the scope and sequence of their investigations. Using resources from Techbook, students determine for themselves which resources they will Explore to answer the Lesson Questions. It is important to note that each student will choose multiple resources, but no one student is expected to use all the resources available. Students also determine the order in which to explore these resources and how to record their findings.

NGSS Components

SEP	CCC
■ Analyzing and Interpreting Data ■ Engaging in Argument from Evidence	■ Systems and System Models

Lesson Question: How Are Photosynthesis and Cellular Respiration Related?

Recommended 35 minutes

- Core Interactive Text: How Are Photosynthesis and Cellular Respiration Related?

TEACHER NOTE Connections: Crosscutting Concept: Systems and System Models: In this concept, students analyze the systems of photosynthesis and of cellular respiration, by defining the boundaries of each, as well as inputs (reactants) and outputs (products). To do so, students use models to simulate the flow of energy, matter, and interactions within cellular systems at different scales. Encourage students to summarize their learning about each system and its components and processes by drawing pictures accompanied by writing a few phrases or a sentence. Access this strategy by clicking the Professional Learning tab on Science Techbook. Click Strategies & Resources, then Spotlight On Strategies (SOS). Journals is found underneath Key Ideas and Details.

- Image: Green Algae
- Image: Molecular Structure of Glucose
- Image: Chlorophyll

TEACHER NOTE Have students record each reactant and product of photosynthesis on index cards, as well as an addition sign and a yield sign. Have them create the photosynthesis reaction on their work surface using these cards. Have them keep these cards so that after they read the next section of text on cellular respiration, they will rearrange the cards to show that reaction.

- Video: Chlorophyll
- Image: Complementary Processes

TEACHER NOTE Have students rearrange their index cards of reactants and products to show the chemical reaction for aerobic cellular respiration. Emphasize to students that the equation for photosynthesis is the reverse of cellular respiration. Ask students to consider why this is important. What would happen if plants or animals produced products that were not consumed by other organisms? Ask them to draw an illustration showing how plants and other organisms recycle the matter used in these two processes?

TEACHER NOTE Misconception: Students may think that photosynthesis occurs only in green plants. Explain that photosynthesis requires a pigment that is able to absorb light. Chlorophyll is the main pigment in plants that absorbs light, and it causes the green color in most plants. However, there are other pigments found in brown or red plants and algae that absorb other wavelengths of light. Photosynthesis also occurs in some Archaea and bacteria.

UNIT 2: Energy, Matter, and Life

TEACHER NOTE Misconception: Students may think that plants take in all needed substances through their roots. Although water and some other minerals are absorbed by the roots, plants take in carbon dioxide through leaves.

TEACHER NOTE Misconception: Students may think that animals only respire and plants only photosynthesize. Although most plants do photosynthesize, all plants, like all animals, respire.

- Video: Comparing Photosynthesis with Cellular Respiration
- Video: Photosynthesis and Cellular Respiration

Formative Assessment:

Throughout Explore, Technology Enhanced Items (TEIs) are embedded as multi-dimensional formative checks for understanding. You can use the data they provide to

- assign additional support
- extend learning
- design additional learning tasks to clarify student misconceptions

The Explore TEIs provide students with three attempts to demonstrate their proficiency. Scaffolded feedback is provided for each attempt. If a student does not achieve proficiency by the third attempt, a media asset is provided as an additional learning opportunity.

TEACHER NOTE Practices: Science and Engineering Practice: Engaging in Argument from Evidence: In this item, students construct and present a written argument about the similarities and differences between cellular respiration and photosynthesis by sorting features of both based on data and evidence about the natural world. These pathways share many common characteristics but evolved independently. Help students think critically about the characteristics of cellular respiration and photosynthesis by using the Myth Bustin'! strategy. Present students with different statements about cellular respiration or photosynthesis and have them cite evidence to refute it. The Myth Bustin'! strategy is found on the Professional Learning tab. Click on Strategies & Resources, then Spotlight On Strategies (SOS). Myth Bustin'! is found underneath Cites Evidence.

TEI Photosynthesis or Respiration?

Lesson Question: How Does the Process of Photosynthesis Work? **Recommended 55 minutes**

- Core Interactive Text: How Does the Process of Photosynthesis Work?
- Exploration: Photosynthesis
- Image: Thylakoids
- Image: Calvin Cycle
- Video: Light-Independent Reaction in Detail
- Video: Storing the Glucose Produced by Photosynthesis
- Video: Chloroplasts
- Image: Chloroplast Structure
- Video: Photosystems

TEACHER NOTE **Practices: Science and Engineering Practice: Analyzing and Interpreting Data:** In this item, students interpret two curves illustrating the differences between carbon dioxide usage and photosynthesis in two species of plants. They analyze data using models in order to make valid and reliable scientific claims. Help them interpret the graph using the Hot Potato strategy. As they examine the data, have them pose questions that pertain to the graph. The Hot Potato strategy is found on the Professional Learning tab. Click on Strategies & Resources, then Spotlight On Strategies (SOS). Hot Potato is found underneath Questioning.

TEI C3 and C4 Species

Explore More Resources

Resources in Explore More Resources support differentiation within your classroom by

- providing additional visualization of content
- affording extension of content to those students ready for acceleration
- offering Lexile reading levels for reading passages

Online explorations and hands-on experiences are provided so that students can conduct virtual investigations, collect and design investigations, and collect and analyze data; these skills are essential to developing scientific understanding

Explain `45–90 minutes`

In Explore, students
1. uncovered scientific understandings
2. conducted investigations
3. analyzed data, text, and other media resources
4. collected evidence to support their scientific explanation

In Explain, provide students with time to formally compose their scientific explanations around the Explain or student-generated questions using evidence collected from Explore.

Scientific explanations are student responses, either written or orally presented, that explain scientific phenomena based upon evidence. Developing a scientific explanation requires students to analyze and interpret data to construct meaning out of the data. There are three main components to the scientific explanation: the claim, the evidence, and the reasoning.

To help students to communicate their scientific explanations, allow them to utilize the multimedia creation tools such as Board Builder and Whiteboard. Remind them that they may upload image, audio, and video files using the "attach file" option to communicate their scientific explanations.

Students may construct their scientific explanations individually or within a small group of students. Students should communicate their explanations with other classmates, and provide constructive criticism and refine their explanations prior to submission to the teacher. If explanations are used as a formative assessment, you can provide additional feedback and comments to support students as they refine their explanations.

EXPLAIN

Can you explain the steps involved in converting sunlight into the energy that is stored in food and compare the processes of cellular respiration and photosynthesis?

Elaborate with STEM `45–180 minutes`

*Elaborate with STEM are optional extension resources available after students have demonstrated proficiency with standards addressed previously in the concept.

NGSS Components

SEP	CCC
■ Asking Questions and Defining Problems ■ Planning and Carrying Out Investigation ■ Using Mathematics and Computational Thinking ■ Constructing Explanations and Designing Solutions ■ Obtaining, Evaluating, and Communicating Information	■ Cause and Effect ■ Structure and Function

STEM In Action `45 minutes`

STEM in Action ties the scientific concepts to real-world applications, with many connecting to STEM careers. Technology Enhanced Items (TEIs) expect students to critically read the Core Interactive Text (CIT) and review the provided media resources.

Applying Photosynthesis

- Core Interactive Text: Applying Photosynthesis
- Image: Red Tides
- Video: Nutrient Pollution
- Video: Better Biofuels
- Image: Artificial Photosynthesis

TEACHER NOTE Use this as a formative assessment to check that students understand where carbon dioxide in the atmosphere comes from and how photosynthesis removes it from the atmosphere. Students could do this individually or in pairs.

Have students share their arguments in small groups and critique one another's reasoning for why there should, or should not, be financing for artificial photosynthesis and biofuels. Instruct students to base their reasoning on evidence learned from the lesson or researched on the Internet. Students should also use data provided by the two graphs to support their arguments.

Inform students that their reports must be structured to include an introduction, supportive paragraphs, and a conclusion, and the connections between concepts and ideas should be cohesive. Introductions should clearly present the student's opinion, and there should be at least one paragraph in the body for every argument or claim that is used. Finally, the conclusion should summarize and restate the student's position.

Have students evaluate the credibility and accuracy of the Internet sources they use for their reports. If necessary, review with students the criteria for credible sources.

TEI Removing Atmospheric Carbon Dioxide

STEM Project Starters

STEM Project Starters provide additional real-world contexts that require students to apply and extend their content knowledge related to the concept. STEM Project Starters can also serve as an alternative instructional hook presented at the beginning of the learning progression. The project can then be revisited throughout and at the end of the 5E learning cycle, for students to apply content knowledge.

STEM Project Starter: Mimicking Photosynthesis   **Recommended 60 minutes**

How can photosynthesis be performed in a laboratory?

TEACHER NOTE This is a good activity for students to work on with a partner or in a small group. It is a good opportunity for students to make connections with the lesson concept and the advance in both technology and design that the study of photosynthesis has brought to a field. Allow students a chance to display their findings in the classroom, particularly if it is not possible to upload their display. Use this as a formative assessment to check that students understand the connection between photosynthesis and biomimicry.

■ Reading Passage: Biomimicry and Photosynthesis

TEI Mimicking Photosynthesis

STEM Project Starter Modeling Carbon Storage
Through Photosynthesis

`Recommended 90 minutes`

How does photosynthesis lead to carbon storage in forests?

TEACHER NOTE This is a good activity for students to work in small groups. It gives students the opportunity to collect data and display it in a graph. Students also have the opportunity to design a model to represent measurements of carbon storage across the United States.

Students may require guidance in choosing the best way to design their model and how to best represent the data. You may advise them of ways to represent different pieces of information through the use of different colors or materials. Students may complete the activity in pairs or small groups. Use this as a summative assessment to show they understand how plants remove carbon dioxide from the atmosphere.

TEI Modeling Carbon Storage through Photosynthesis

- Activity: Engineering Design Sheet

STEM Project Starter Fertilizers and Plant Growth

`Recommended 90 minutes`

How do fertilizers affect aquatic plants?

TEI Fertilizers and Plant Growth

TEACHER NOTE This is a good activity for students to work in small groups. It is an opportunity to construct a testable question, design an experiment, and carry it out for analysis. Students can apply what they learned about photosynthesis as they carry out this experiment on aquatic plant growth. An experiment with aquatic plants can be easily carried out in a classroom, but students may need help gathering materials. Students may then complete the activity in pairs or small groups. Use this as a formative assessment to check that students understand the effects of fertilizers on plant growth.

Evaluate `45–90 minutes`

Explain Question:
Can you explain the steps involved in converting sunlight into the energy that is stored in food and compare the processes of cellular respiration and photosynthesis?

Lesson Questions (LQ):
- How are photosynthesis and cellular respiration related?
- How does the process of photosynthesis work?

Throughout instruction and the 5E learning cycle, you will have collected formative assessment data to drive the assignment of resources and experiences to students. Evaluate is intended to include summative assessment checks for proficiency. You can use the Explain and Lesson Questions for the concept as a summative assessment in a variety of ways such as these:

- Post each Lesson Question (LQ) in various locations in the classroom, and have small groups of students generate claim statements related to the Lesson Question (LQ). Other students can add to the claim, or refute the claim, during a gallery walk where they place additional pieces of evidence on each Lesson Question (LQ) poster.
- Assign small groups of students to each Lesson Question (LQ) and have the groups generate a poster, board, graphic, or piece of text that answers the question. Use a jigsaw approach and create a second set of groups that contain members from each Lesson Question (LQ) group to share their ideas.
- Ask students to return to their initial ideas for the Explain question and add additional details and evidence.

Encourage students to review the concept review and complete the Student Self-Check practice assessment prior to assigning the Summative Teacher Concept assessment.

- Student Review and Practice Assessment
- Teacher Concept Assessments

Cellular Respiration

The Five E Instructional Model

Science Techbook follows the 5E instructional model. This Model Lesson includes strategies for each of the 5Es. As you design the inquiry-based learning experience for students, be sure to collect data during instruction to drive your instructional decisions. Point-of-use teacher notes are also provided within each E-tab.

Engage   45–90 minutes

Engage Media Resources

The resources found in Engage are intended to stimulate students by exposing them to a phenomenon relevant to the content of the lesson. Engage also provides examples of relevant real-world applications that allow students to begin to make observations and relate the science content to their everyday lives. The Core Interactive Text (CIT) and media resources are carefully designed to prompt students to begin asking questions that they can investigate during the Explore phase of the lesson. They should also start collecting evidence to address the Explain question located at the bottom of the Engage page.

> **TEACHER NOTE** **Investigative Phenomenon:** Free running is an emerging sport that requires a great deal of expertise and the expenditure of a vast amount of energy. Use the text in the first paragraph and the video, Free Running Is Not Free, to get students to consider the origin of energy needed for such a vigorous activity (They will have some ideas from earlier lessons.) and how this energy could be liberated. The analogy of food as a form of fuel is a useful segue to get them thinking about how the process of "burning" food could take place safely inside a living organism

- Core Interactive Text: Fueling Fun
- Video: Free Running Is Not Free
- Video: Introducing Cellular Respiration

Explain Question

The Explain question focuses students on gathering information in the Explore section. The Explain question can be used to

- Record what students already know related to the Explain question.
- Serve as a template or model for students to generate their own scientific questions.
- Collect evidence as students work through the lesson.
- Allow students to reflect on their growth before and after the lesson.

Explain

Explain the steps involved in converting food into the energy needed to move these athletes.

- Image: They're Off...

Engage Formative Assessment

Technology Enhanced Items (TEIs) found on the Engage page enable you to collect data on students' prior knowledge and identify the common misconceptions they may possess that are related to the topic of study. These items are designed as quick checks for understanding and allow each student one attempt at each question. You can use the data collected to decide whether to assign additional resources to the class, or determine what individual or groups of students may need reinforcement or accelerated learning, prior to completing the Explore portion of the lesson.

TEACHER NOTE Use this student response to formatively evaluate students' prior knowledge of the concept. The Model Lesson provides information on common student misconceptions. Students should compare the answer they have given here to the one they give in Explain.

TEI Your Ideas

Before You Begin

What Do I Already Know about Cellular Respiration?

TEACHER NOTE This item is a formative pre-assessment intended to provide the teacher with feedback on prior knowledge of this topic. In middle school, students should have learned that respiration has reactants and products and should have an understanding of what these are and how they are related. Use as a think-pair-share activity.

TEI Reactants and Products

TEACHER NOTE Use this as a formative pre-test of these common student misconceptions. Students selecting A are probably incorrectly equating feeling hot as being caused by sweating. Students selecting B probably have the misconception that respiration is a form of burning or combustion. Students selecting D may think that cells only respire when used and are in use.

TEI Core Temperature

TEACHER NOTE This activity is a formative pre-assessment intended to provide the teacher with feedback on prior students' knowledge of this topic. Its main focus is to determine whether students are aware that cellular respiration is a multi-staged process and whether they are familiar with any of the steps or organelles involved with that process. Consider having students work in pairs on this pre-assessment.

TEI **Describing Respiration**

TEACHER NOTE Students may have heard of ATP, and some may have an understanding of its role as a high-energy compound. This formative assessment will provide some feedback on their understanding of how ATP works. Use as a think-pair-share activity.

TEI **ATP**

- Video: Cell Structure and Organelles
- Video: ATP
- Video: Chemical Reactions

Explore `100 minutes`

Lesson Questions (LQs):
1. What are the steps involved in cellular respiration?
2. What are the differences between aerobic and anaerobic respiration?

Effective science instruction involves a student-centered rather than a teacher-centered approach. This can be accomplished either with Directed Inquiry or Guided Inquiry, depending on the needs and abilities of your class. Encourage students to select a variety of resources in their pursuit of answers as they work through Explore, with the end goal of constructing their scientific explanation in the Explain tab.

Directed Inquiry	Guided Inquiry
In Directed Inquiry, teachers provide students with a sequence of specified resources, challenging questions, and clear outcomes. Within this context students are given the opportunity to interact independently with each resource as prescribed by the teacher. Often different students groups can be guided through several different resources at the same time. For example, one group could work on a reading passage while a second group conducts a small-group Hands-On Activity with the teacher, and a third group is independently engaged with an online interactive resource.	In Guided Inquiry, students have independence to decide the scope and sequence of their investigations. Using resources from Techbook, students determine for themselves which resources they will Explore to answer the Lesson Questions. It is important to note that each student will choose multiple resources, but no one student is expected to use all the resources available. Students also determine the order in which to explore these resources and how to record their findings.

NGSS Components

SEP	CCC
■ Developing and Using Models ■ Planning and Carrying Out Investigations ■ Analyzing and Interpreting Data ■ Constructing Explanations and Designing Solutions	■ Systems and System Models ■ Energy and Matter: Flows, Cycles, and Conservation

Lesson Question: What Are the Steps Involved in Cellular Respiration? `Recommended 50 minutes`

TEACHER NOTE Connections: Crosscutting Concept: Energy and Matter: Flows, Cycles, and Conservation: In this concept, students describe changes of energy and matter in living things in terms of energy and matter flowing into, out of, and within their systems through the process of cellular respiration. They learn that the total amount of energy is conserved during cellular respiration because energy is transformed, not gained or lost.

TEACHER NOTE Misconception: Students may think that respiration only refers to breathing. In fact, cellular respiration involves a series of chemical reactions that take place inside cells and yield usable energy from complex organic molecules.

- Core Interactive Text: What Are the Steps Involved in Cellular Respiration?
- Video: Introducing Cellular Respiration
- Video: What Is ATP?

TEACHER NOTE Practices: Science and Engineering Practice: Developing and Using Models: In this exploration, students will use a model experimental set-up to generate data to support the explanation of the conversion of oxygen to carbon dioxide during cellular respiration. After students have completed the exploration, expand on student understanding by leading a class discussion regarding why physical exercise results in heavier breathing.

- Exploration: Cellular Respiration
- Video: Anaerobic Respiration
- Image: Glycolysis
- Video: Glycolysis and Cellular Respiration
- Video: Krebs Cycle
- Image: Krebs Cycle
- Video: Electron Transport Chain
- Image: Electron Transport Chain
- Video: The Electron Transport Chain as Part of Cellular Respiration

Formative Assessment:

Throughout Explore, Technology Enhanced Items (TEIs) are embedded as multi-dimensional formative checks for understanding. You can use the data they provide to

- assign additional support
- extend learning
- design additional learning tasks to clarify student misconceptions

The Explore TEIs provide students with three attempts to demonstrate their proficiency. Scaffolded feedback is provided for each attempt. If a student does not achieve proficiency by the third attempt, a media asset is provided as an additional learning opportunity.

TEACHER NOTE **Connections: Crosscutting Concept: Systems and System Models:** Students will analyze the system of chemical reactions that constitute cellular respiration, defining the initial conditions and boundaries of the organelles involved in the process and the inputs and outputs (reactants and products) of each stage. After students complete this item, they can create diagrams of a cell to model the steps of cellular respiration, writing the reactants and products of each stage on the structures of the cell in which they take place.

TEI Cellular Respiration

Lesson Question: What Are the Differences Between Aerobic and Anaerobic Respiration?

Recommended 50 minutes

- Core Interactive Text: What Are the Differences Between Aerobic and Anaerobic Respiration?
- Image: Beta Haemolytic Streptococci Bacteria
- Video: Anaerobic Respiration
- Video: Comparing Aerobic and Anaerobic Respiration

TEACHER NOTE **Practices: Science and Engineering Practice: Analyzing and Interpreting Data:** In this activity, students analyze data from an investigation of yeast fermentation by making qualitative and quantitative measures and graphing their results to make valid and reliable claims about cellular respiration. In the analysis and conclusion at the end of the lab, students consider the limitations of their data analysis and suggest ways in which they could collect more accurate data. If time permits, students can redo the investigation using the improved data collection methods they suggest and revise their conclusions in light of the new data.

- Hands-On Activity: Yeast Fermentation

TEACHER NOTE **Practices: Science and Engineering Practice: Planning and Carrying Out Investigations:** In the Hands-On Activity that follows, students plan an investigation collaboratively to produce data to serve as the basis for evidence to support explanations of aerobic and anaerobic respiration. Students perform two investigations, manipulating a variable in the second investigation, and graph the data they collect. To extend this activity, instruct groups to design and, if time permits, conduct a third investigation in which they manipulate the variable again. Students can then revise or expand their explanations based on the new data they gather.

- Hands-On Activity: Aerobic vs. Anaerobic Respiration
- Image: Fermentation
- Video: Fermentation

TEACHER NOTE Practices: Science and Engineering Practice: Constructing Explanations and Designing Solutions: Students apply scientific ideas and evidence to provide an explanation of sauerkraut creation and the relationship between cellular respiration and the amount of oxygen available to cells. To extend this item, assign students a food that requires fermentation to prepare (or allow them to choose one) and instruct them to create a poster or presentation explaining the processes by which the food is created. Students should cite evidence for the chemical reactions that enable the creation of the food.

TEI Oxygen and Cellular Respiration

TEI Making Sauerkraut

Explore More Resources

Resources in Explore More Resources support differentiation within your classroom by

- providing additional visualization of content
- affording extension of content to those students ready for acceleration
- offering Lexile reading levels for reading passages

Online explorations and hands-on experiences are provided so that students can conduct virtual investigations, collect and design investigations, and collect and analyze data; these skills are essential to developing scientific understanding.

Explain `45–90 minutes`

In Explore, students
1. uncovered scientific understandings
2. conducted investigations
3. analyzed data, text, and other media resources
4. collected evidence to support their scientific explanation

In Explain, provide students with time to formally compose their scientific explanations around the Explain or student-generated questions using evidence collected from Explore.

Scientific explanations are student responses, either written or orally presented, that explain scientific phenomena based upon evidence. Developing a scientific explanation requires students to analyze and interpret data to construct meaning out of the data. There are three main components to the scientific explanation: the claim, the evidence, and the reasoning.

To help students to communicate their scientific explanations, allow them to utilize the multimedia creation tools such as Board Builder and Whiteboard. Remind them that they may upload image, audio, and video files using the "attach file" option to communicate their scientific explanations.

Students may construct their scientific explanations individually or within a small group of students. Students should communicate their explanations with other classmates, and provide constructive criticism and refine their explanations prior to submission to the teacher. If explanations are used as a formative assessment, you can provide additional feedback and comments to support students as they refine their explanations.

EXPLAIN

Explain the steps involved in converting food into the energy needed to move these athletes.

Elaborate with STEM `45–135 minutes`

*Elaborate with STEM are optional extension resources available after students have demonstrated proficiency with standards addressed previously in the concept.

NGSS Components

SEP	CCC
■ Asking Questions and Defining Problems ■ Planning and Carrying Out Investigation ■ Using Mathematics and Computational Thinking ■ Constructing Explanations and Designing Solutions ■ Obtaining, Evaluating, and Communicating Information	■ Cause and Effect ■ Structure and Function

STEM In Action `45 minutes`

STEM in Action ties the scientific concepts to real-world applications, with many connecting to STEM careers. Technology Enhanced Items (TEIs) expect students to critically read the Core Interactive Text (CIT) and review the provided media resources.

Applying Cellular Respiration

- Core Interactive Text: Applying Cellular Respiration
- Image: Cramps
- Video: Maximizing Power Output
- Video: The Chemistry of Making Bread
- Video: Ethanol
- Video: Fermentation
- Image: Fermentation Vats

TEACHER NOTE Use this as a formative assessment to determine whether students can make connections between the process of fermentation and the production of ethanol to the wider economy, oil production technology, and world food supplies.

TEI Fermenting Fuel

STEM Project Starters

STEM Project Starters provide additional real-world contexts that require students to apply and extend their content knowledge related to the concept. STEM Project Starters can also serve as an alternative instructional hook presented at the beginning of the learning progression. The project can then be revisited throughout and at the end of the 5E learning cycle, for students to apply content knowledge.

STEM Project Starter: Just Plain Nuts

Recommended 60 minutes

Can you compare the energy in different types of nuts?

> **TEACHER NOTE** Students should submit their reports at least twice—once when they have designed their procedure and once when they have conducted the investigation. Consider using this as a summative assessment.

TEI Just Plain Nuts

STEM Project Starter: Measuring Cellular Respiration
Virtual Lab

Recommended 30 minutes

How can carbon dioxide be measured?

> **TEACHER NOTE** Refer to the Cellular Respiration Teacher Guide. A Student Guide version is also available. This project provides a summative assessment.

■ Exploration: Cellular Respiration

TEI Cellular Respiration

STEM Project Starter: Bread Making

Recommended 30 minutes

What is the science behind making bread?

> **TEACHER NOTE** This formative STEM project relates the respiration of yeast to the food technology of baking. Consider having the students investigate the nutritional advantages of bread baked with yeast over bread made with baking powder. Use the students' submissions as the basis for a class discussion and relate it to careers in bakeries.

TEI Explaining My Recipe

Evaluate `45–90 minutes`

Explain Question:
Explain the steps involved in converting food into the energy needed to move these athletes.

Lesson Questions (LQ):
- What are the steps involved in cellular respiration?
- What are the differences between aerobic and anaerobic respiration?

Throughout instruction and the 5E learning cycle, you will have collected formative assessment data to drive the assignment of resources and experiences to students. Evaluate is intended to include summative assessment checks for proficiency. You can use the Explain and Lesson Questions for the concept as a summative assessment in a variety of ways such as these:

- Post each Lesson Question (LQ) in various locations in the classroom, and have small groups of students generate claim statements related to the Lesson Question (LQ). Other students can add to the claim, or refute the claim, during a gallery walk where they place additional pieces of evidence on each Lesson Question (LQ) poster.
- Assign small groups of students to each Lesson Question (LQ) and have the groups generate a poster, board, graphic, or piece of text that answers the question. Use a jigsaw approach and create a second set of groups that contain members from each Lesson Question (LQ) group to share their ideas.
- Ask students to return to their initial ideas for the Explain question and add additional details and evidence.

Encourage students to review the concept review and complete the Student Self-Check practice assessment prior to assigning the Summative Teacher Concept assessment.

- Student Review and Practice Assessment
- Teacher Concept Assessments

Nutrient Cycles

The Five E Instructional Model

Science Techbook follows the 5E instructional model. This Model Lesson includes strategies for each of the 5Es. As you design the inquiry-based learning experience for students, be sure to collect data during instruction to drive your instructional decisions. Point-of-use teacher notes are also provided within each E-tab.

Engage   45–90 minutes

Engage Media Resources

The resources found in Engage are intended to stimulate students by exposing them to a phenomenon relevant to the content of the lesson. Engage also provides examples of relevant real-world applications that allow students to begin to make observations and relate the science content to their everyday lives. The Core Interactive Text (CIT) and media resources are carefully designed to prompt students to begin asking questions that they can investigate during the Explore phase of the lesson. They should also start collecting evidence to address the Explain question located at the bottom of the Engage page.

> **TEACHER NOTE** **Investigative Phenomenon:** The recycling of resources is an essential activity in a sustainable society. Students will be familiar with the practice of recycling certain items—paper, bottles and cans—but are probably unaware of what happens to them once they enter the recycling system. Similarly, they may be aware that nature recycles water and important nutrients, but have no knowledge of the processes that occur to accomplish this recycling. Use the text, images, and videos in Engage to pique students' interest in the systems behind recycling trash. Were students aware that recycling trash was such a complex process involving different pathways? Do they think all the trash is recycled? Is energy used in the recycling process (and if so where does it come from)? When students have considered these questions, have them extend their thinking to what they know about natural systems. Have students work in pairs to produce a T-chart that compares recycling of trash to what they know about nature's recycling systems.

- Core Interactive Text: Recycle and Reuse
- Image: Recycling Resources
- Video: Recycling

Explain Question

The Explain question focuses students on gathering information in the Explore section. The Explain question can be used to

- Record what students already know related to the Explain question.
- Serve as a template or model for students to generate their own scientific questions.
- Collect evidence as students work through the lesson.
- Allow students to reflect on their growth before and after the lesson.

Explain

How do water and nutrients cycle through the environment?

> - Image: Nutrient Recycling in Action

Engage Formative Assessment

Technology Enhanced Items (TEIs) found on the Engage page enable you to collect data on students' prior knowledge and identify the common misconceptions they may possess that are related to the topic of study. These items are designed as quick checks for understanding and allow each student one attempt at each question. You can use the data collected to decide whether to assign additional resources to the class, or determine what individual or groups of students may need reinforcement or accelerated learning, prior to completing the Explore portion of the lesson.

> **TEACHER NOTE** Use this formative question to evaluate students' prior knowledge of cellular respiration and its role in carbon cycling. Students can pair up to discuss and agree on answers before sharing with the entire class.

TEI **Breaking It Down**

Before You Begin

What Do I Already Know about Nutrient Cycles?

> **TEACHER NOTE** This formative activity is intended to provide the teacher with feedback on student misconceptions of cellular respiration and the carbon cycle. Students can work individually or in pairs (think-pair-share) to come to consensus.

TEI **Contributing to the Carbon Cycle**

TEACHER NOTE This formative activity is intended to provide the teacher with feedback on student misconceptions of nutrients and to gauge prior knowledge of environmental problems due to excess nutrients. This question is best approached in a class discussion format, possibly with one half of the class taking one position (no such thing as too many nutrients) and the other half taking the opposite position.

TEI Nutritious Nutrients?

TEACHER NOTE This formative activity is intended to provide the teacher with feedback on prior knowledge of water's physical properties. Students can work in pairs to corroborate opinions before sharing with the entire class.

TEI Wonderful Water

TEACHER NOTE This formative activity is intended to provide the teacher with feedback on prior knowledge of organic and inorganic compounds. Students can work in pairs to corroborate opinions before sharing with the entire class.

TEI Organic versus Inorganic?

- Video: Compounds, Elements, and Atoms
- Video: Photosynthesis and Cellular Respiration
- Video: Chemical Reactions
- Video: Recycling Aluminum
- Video: Shopping Carts to Manhole Covers

Explore `135 minutes`

Lesson Questions (LQs):
- What are the processes and features of the water cycle?
- What are the processes and features of the carbon cycle?
- What are the processes and features of the nitrogen cycle?
- What are the processes and features of the phosphorus cycle?

Effective science instruction involves a student-centered rather than a teacher-centered approach. This can be accomplished either with Directed Inquiry or Guided Inquiry, depending on the needs and abilities of your class. Encourage students to select a variety of resources in their pursuit of answers as they work through Explore, with the end goal of constructing their scientific explanation in the Explain tab.

Directed Inquiry	Guided Inquiry
In Directed Inquiry, teachers provide students with a sequence of specified resources, challenging questions, and clear outcomes. Within this context students are given the opportunity to interact independently with each resource as prescribed by the teacher. Often different students groups can be guided through several different resources at the same time. For example, one group could work on a reading passage while a second group conducts a small-group Hands-On Activity with the teacher, and a third group is independently engaged with an online interactive resource.	In Guided Inquiry, students have independence to decide the scope and sequence of their investigations. Using resources from Techbook, students determine for themselves which resources they will Explore to answer the Lesson Questions. It is important to note that each student will choose multiple resources, but no one student is expected to use all the resources available. Students also determine the order in which to explore these resources and how to record their findings.

NGSS Components

SEP	CCC
■ Developing and Using Models ■ Analyzing and Interpreting Data	■ Systems and System Models ■ Energy and Matter: Flows, Cycles, and Conservation

Lesson Question: What Are the Processes and Features of the Water Cycle?

Recommended 30 minutes

TEACHER NOTE Connections: Crosscutting Concept: Energy and Matter: Flows, Cycles, and Conservation: In this concept, students will study the flow of matter through cycles. They learn that matter cannot be created or destroyed. It only moves between one place and another place. Help students see that matter can travel through different paths in the cycle by drawing flowcharts on the board. Fill in the flowcharts as students study each possible step of the cycle.

- Core Interactive Text: What Are the Processes and Features of the Water Cycle?

TEACHER NOTE Students encounter the water cycle every day. Use the Connect the Dots strategy to help them identify common parts of the water cycle. Ask them where water goes and where it comes from. Where does water go after a shower? Where does water go when it is boiled? Where does water come from when it condenses on a car windshield? Where does tap water originate? The Connect the Dots strategy is found on the Professional Learning tab. Click on Strategies & Resources, then Spotlight On Strategies (SOS). Connect the Dots is found under Research.

- Image: The Water Cycle
- Video: Deltas

Formative Assessment:

Throughout Explore, Technology Enhanced Items (TEIs) are embedded as multi-dimensional formative checks for understanding. You can use the data they provide to

- assign additional support
- extend learning
- design additional learning tasks to clarify student misconceptions

The Explore TEIs provide students with three attempts to demonstrate their proficiency. Scaffolded feedback is provided for each attempt. If a student does not achieve proficiency by the third attempt, a media asset is provided as an additional learning opportunity.

TEACHER NOTE Practices: Science and Engineering Practice: Developing and Using Models: This item, Drought, requires students to apply models of the water cycle to explain why disruptions in the water cycle are harmful. They use a model based on evidence to illustrate and predict the relationships between components of a system. Help them understand the significance of the water cycle by using the Pecha Kucha strategy. Provide students different topics on different drought and flooding events for them to research. The Pecha Kucha strategy is found on the Professional Learning tab. Click on Strategies & Resources, then Spotlight On Strategies (SOS). Pecha Kucha is found under Research.

TEI Drought

Lesson Question: What Are the Processes and Features of the Carbon Cycle?

Recommended 50 minutes

TEACHER NOTE **Misconception:** Students may think that the carbon cycle consists only of photosynthesis and respiration, and that only animals carry out cellular respiration. The carbon cycle involves many processes in addition to photosynthesis and cellular respiration, such as combustion and decomposition. Organisms such as plants undergo both photosynthesis and cellular respiration.

- Core Interactive Text: What Are the Processes and Features of the Carbon Cycle?
- Video: The Carbon Cycle
- Assignment: Carbon Cycle
- Video: Science in Progress: Biogeochemical Cycles and the Ocean

TEACHER NOTE **Connections: Crosscutting Concept: Systems and System Models:** This item, Carbon Cycle, requires students to create a model of the carbon cycle by tracking the movement of a carbon atom through the cycle. They use models to simulate the flow of matter within and between systems at different scales. Help students see how carbon moves through the environment by using the Paper Slide strategy. Have them create a story about the movement of carbon in the environment. The Paper Slide strategy is found on the Professional Learning tab. Click on Strategies & Resources, then Spotlight On Strategies (SOS). Paper Slide is found under Research.

TEI Carbon Cycle

Lesson Question: What Are the Processes and Features of the Nitrogen Cycle?

Recommended 20 minutes

TEACHER NOTE **Misconception:** Students may think that because nitrogen and phosphorus are nutrients, adding more to the environment must always be beneficial to the environment. In excess, nitrogen and phosphorus, as well as atmospheric carbon, can disrupt the flow of nutrient cycles and cause harm to living things.

- Core Interactive Text: What Are the Processes and Features of the Nitrogen Cycle?
- Video: Nitrogen-Fixing Bacteria
- Video: Science in Progress: Reviewing Nutrient Cycles

Lesson Question: What Are the Processes and Features of the Phosphorus Cycle?

`Recommended 35 minutes`

- Core Interactive Text: What Are the Processes and Features of the Phosphorus Cycle?
- Exploration: Nutrient Cycles
- Video: The Phosphorus Cycle
- Image: Phosphorus

TEACHER NOTE Practices: Science and Engineering Practice: Analyzing and Interpreting Data: In this item, students will think about how technology can be used to improve farming. They analyze data using tools in order to determine an optimal design solution. Before students complete this item, help them study how fertilizers flow from farms by demonstrating a model of how water moves from farmland. Place dirt in a box and pour water over it. Ask them where the water goes. What strategies prevent water carrying phosphate from flowing away from the dirt?

TEI Feeding Plants

Explore More Resources

Resources in Explore More Resources support differentiation within your classroom by

- providing additional visualization of content
- affording extension of content to those students ready for acceleration
- offering Lexile reading levels for reading passages

Online explorations and hands-on experiences are provided so that students can conduct virtual investigations, collect and design investigations, and collect and analyze data; these skills are essential to developing scientific understanding.

Explain `45–90 minutes`

In Explore, students
1. uncovered scientific understandings
2. conducted investigations
3. analyzed data, text, and other media resources
4. collected evidence to support their scientific explanation

In Explain, provide students with time to formally compose their scientific explanations around the Explain or student-generated questions using evidence collected from Explore.

Scientific explanations are student responses, either written or orally presented, that explain scientific phenomena based upon evidence. Developing a scientific explanation requires students to analyze and interpret data to construct meaning out of the data. There are three main components to the scientific explanation: the claim, the evidence, and the reasoning.

To help students to communicate their scientific explanations, allow them to utilize the multimedia creation tools such as Board Builder and Whiteboard. Remind them that they may upload image, audio, and video files using the "attach file" option to communicate their scientific explanations.

Students may construct their scientific explanations individually or within a small group of students. Students should communicate their explanations with other classmates, and provide constructive criticism and refine their explanations prior to submission to the teacher. If explanations are used as a formative assessment, you can provide additional feedback and comments to support students as they refine their explanations.

EXPLAIN
How do water and nutrients cycle through the environment?

Elaborate with STEM   45–130 minutes

*Elaborate with STEM are optional extension resources available after students have demonstrated proficiency with standards addressed previously in the concept.

NGSS Components

SEP	CCC
■ Obtaining, Evaluating, and Communicating Information ■ Asking Questions and Defining Problems ■ Planning and Carrying Out Investigation ■ Analyzing and Interpreting Data ■ Using Mathematics and Computational Thinking ■ Constructing Explanations and Designing Solutions ■ Engaging in Argument from Evidence	■ Stability and Change ■ Cause and Effect ■ Structure and Function ■ Systems and System Models ■ Patterns ■ Energy and Matter: Flows, Cycles, and Conservation ■ Scale, Proportion, and Quantity

STEM In Action   45 minutes

STEM in Action ties the scientific concepts to real-world applications, with many connecting to STEM careers. Technology Enhanced Items (TEIs) expect students to critically read the Core Interactive Text (CIT) and review the provided media resources.

Applying Nutrient Cycles

- Core Interactive Text: Applying Nutrient Cycles
- Video: Industrial Farms
- Video: The Dangers of Nitrogen
- Video: Engineering Biofuel
- Video: Ethanol
- Video: Water Pollution
- Video: Greenhouse Gases Experiment
- Video: Greenhouse Gases Results

TEACHER NOTE This formative assessment will be used to gauge understanding of denitrifying bioreactors and the different types of materials that could be used. Students could work in pairs to interpret the figure and share their answers with the class.

TEI Bioreactor Materials

STEM Project Starters

STEM Project Starters provide additional real-world contexts that require students to apply and extend their content knowledge related to the concept. STEM Project Starters can also serve as an alternative instructional hook presented at the beginning of the learning progression. The project can then be revisited throughout and at the end of the 5E learning cycle, for students to apply content knowledge.

STEM Project Starter Reducing Runoff `Recommended 90 minutes`

How can nutrient pollution be mitigated from agricultural fields?

> **TEACHER NOTE** This summative project will provide students the opportunity to apply what they have learned about nutrient cycles to the problem of nutrient pollution. Students can work in pairs, or this can be formatted as a class project and discussion.

TEI Reducing Runoff

STEM Project Starter Contrasting Fuel Types `Recommended 60 minutes`

How do different fuel sources affect the carbon cycle?

> **TEACHER NOTE** This summative project will provide students the opportunity to apply what they have learned about the carbon cycle to different types of fuels. Students can work in pairs, then share results with the class.

TEI Contrasting Fuel Types
TEI Evaluating Costs to Ecosystems

STEM Project Starter Offsetting Emissions `Recommended 90 minutes`

What can people do to ameliorate their carbon emissions?

> **TEACHER NOTE** This summative project will provide students the opportunity to apply what they have learned about the carbon cycle to different methods of offsetting carbon emissions. Students can work in pairs, then share results with the class.

TEI Offsetting Emissions

Evaluate `45–90 minutes`

Explain Question:
How do water and nutrients cycle through the environment?

Lesson Questions (LQ):
- What are the processes and features of the water cycle?
- What are the processes and features of the carbon cycle?
- What are the processes and features of the nitrogen cycle?
- What are the processes and features of the phosphorus cycle?

Throughout instruction and the 5E learning cycle, you will have collected formative assessment data to drive the assignment of resources and experiences to students. Evaluate is intended to include summative assessment checks for proficiency. You can use the Explain question and Lesson Questions for the concept as a summative assessment in a variety of ways such as these:

- Post each Lesson Question (LQ) in various locations in the classroom, and have small groups of students generate claim statements related to the Lesson Question (LQ). Other students can add to the claim, or refute the claim, during a gallery walk where they place additional pieces of evidence on each Lesson Question (LQ) poster.
- Assign small groups of students to each Lesson Question (LQ) and have the groups generate a poster, board, graphic, or piece of text that answers the question. Use a jigsaw approach and create a second set of groups that contain members from each Lesson Question (LQ) group to share their ideas.
- Ask students to return to their initial ideas for the Explain question and add additional details and evidence.

Encourage students to review the concept review and complete the Student Self-Check practice assessment prior to assigning the Summative Teacher Concept assessment.

- Student Review and Practice Assessment
- Teacher Concept Assessments

Mechanical and Chemical Weathering

The Five E Instructional Model

Science Techbook follows the 5E instructional model. This Model Lesson includes strategies for each of the 5Es. As you design the inquiry-based learning experience for students, be sure to collect data during instruction to drive your instructional decisions. Point-of-use teacher notes are also provided within each E-tab.

Engage   45–90 minutes

Engage Media Resources

The resources found in Engage are intended to stimulate students by exposing them to a phenomenon relevant to the content of the lesson. Engage also provides examples of relevant real-world applications that allow students to begin to make observations and relate the science content to their everyday lives. The Core Interactive Text (CIT) and media resources are carefully designed to prompt students to begin asking questions that they can investigate during the Explore phase of the lesson. They should also start collecting evidence to address the Explain question located at the bottom of the Engage page.

> **TEACHER NOTE** **Investigative Phenomenon:** In Engage, students are presented with the image of a landform (a butte). Ask students to closely examine the image. They should select the large image and use the + magnifier tool to blow it up further. Have them work in small groups and suggest and record answers to the following questions—where possible, they should support their answers with evidence from the picture. Do they think the landform has always appeared as it does in the picture? How do they think the rocks that it is made from formed? How do they think the landform itself was formed? Why do they think it is higher than the surrounding rocks? What do they notice about the shape of its rocks compared with those of the beds upon which it stands? How can they account for these differences? What evidence is there of weathering taking place? What evidence (if any) is there of erosion? If yes, what types of erosion may be involved? Choose groups to share their answers. Encourage constructive criticism of their answers. Use these questions repeatedly as students work through Engage and Explore and encounter images of other landforms.

- Core Interactive Text: Observing Mechanical and Chemical Weathering
- Image: Butte Weathering
- Video: The Forces of Wind & Water: Weathering and Erosion
- Video: Water and Weathering
- Image: Mudstone Showing Weathering
- Image: Badlands Shaped by Weathering and Erosion

Explain Question

The Explain question focuses students on gathering information in the Explore section. The Explain question can be used to

- Record what students already know related to the Explain question.
- Serve as a template or model for students to generate their own scientific questions.
- Collect evidence as students work through the lesson.
- Allow students to reflect on their growth before and after the lesson.

Explain

What effect does weathering have on the surface of landforms such as mountains, buttes, mesas, and plateaus?

- Image: Monument Valley

Engage Formative Assessment

Technology Enhanced Items (TEIs) found on the Engage page enable you to collect data on students' prior knowledge and identify the common misconceptions they may possess that are related to the topic of study. These items are designed as quick checks for understanding and allow each student one attempt at each question. You can use the data collected to decide whether to assign additional resources to the class, or determine what individual or groups of students may need reinforcement or accelerated learning, prior to completing the Explore portion of the lesson.

TEACHER NOTE Use this formative assessment to evaluate students' prior knowledge of the concept. This assesses common misconceptions about mechanical weathering, such as that it always appears through machine-driven processes, when actually the freezing and flowing of water and the growing of plant roots can cause mechanical weathering. Incorrect answers: 3 represents a cave formed through chemical weathering, and 4 represents an example of chemical weathering. Students may complete individually or as a think-pair-share.

TEI Mechanical Weathering

Before You Begin
What Do I Already Know about Weathering?

TEACHER NOTE This formative assessment is designed to provide the teacher with feedback about a student's general understanding of weathering and the type of changes it causes to the surface of Earth.

1. Demonstrates erosion more than weathering
2. Addresses a common misconception that mechanical weathering involves machines or tools
3. Changes in temperature can play a role in weathering
4. Wind blowing leaves onto rocks is not an accurate description of weathering

Students should complete individually as homework or as an opening activity, and then discuss in class.

TEI Examples of Weathering

TEACHER NOTE This formative assessment is intended to provide the teacher with feedback on prior knowledge of this topic. Its main focus is to determine whether students are aware of some of the important terms and elements that play a role in the weathering process. Students may confuse terms such as weathering and erosion. While the two often work in conjunction, they are the same thing.

Teachers may opt to have students complete this activity individually as a homework assignment or work together in pairs or small groups to match the definitions.

TEI Elements of Weathering

TEACHER NOTE This formative assessment is intended to provide the teacher with feedback on a student's understanding of the relationship of erosion to weathering. Students often confuse weathering with erosion or think the two work independently of one another, but the two processes actually work together. Teachers may have students complete as homework or answer the question individually, and then discuss as a class.

TEI Weathering and Erosion

- Video: Physical and Chemical Changes in Matter
- Video: Changes in Matter

Explore `135 minutes`

Lesson Questions (LQs):
1. How do mechanical and chemical weathering change Earth's surface?

Effective science instruction involves a student-centered rather than a teacher-centered approach. This can be accomplished either with Directed Inquiry or Guided Inquiry, depending on the needs and abilities of your class. Encourage students to select a variety of resources in their pursuit of answers as they work through Explore, with the end goal of constructing their scientific explanation in the Explain tab.

Directed Inquiry	Guided Inquiry
In Directed Inquiry, teachers provide students with a sequence of specified resources, challenging questions, and clear outcomes. Within this context students are given the opportunity to interact independently with each resource as prescribed by the teacher. Often different students groups can be guided through several different resources at the same time. For example, one group could work on a reading passage while a second group conducts a small-group Hands-On Activity with the teacher, and a third group is independently engaged with an online interactive resource.	In Guided Inquiry, students have independence to decide the scope and sequence of their investigations. Using resources from Techbook, students determine for themselves which resources they will Explore to answer the Lesson Questions. It is important to note that each student will choose multiple resources, but no one student is expected to use all the resources available. Students also determine the order in which to explore these resources and how to record their findings.

NGSS Components

SEP	CCC
■ Planning and Carrying Out Investigations ■ Constructing Explanations and Designing Solutions	■ Cause and Effect ■ Stability and Change

Lesson Question: How Do Mechanical and Chemical Weathering Change Earth's Surface? `Recommended 45 minutes`

TEACHER NOTE Practices: Science and Engineering Practice: Constructing Explanations and Designing Solutions: Throughout this concept, students apply scientific ideas, principles, and evidence to provide an explanation for the phenomena of mechanical and chemical weathering. To introduce this concept, show students the images Abrasion and Exfoliation (both images are found on this page and in Explore More Resources). You can use the Stem Completion strategy to guide students to hypothesize how the geologic formations in these images were formed, including the forces that formed them and the time period over which these changes occurred. This strategy is found on the Professional Learning tab. Click on Strategies & Resources, then Spotlight On Strategies (SOS). Stem Completion is found underneath Instructional Hook.

- Core Interactive Text: How Do Mechanical and Chemical Weathering Change Earth's Surface?
- Image: Abrasion
- Image: Exfoliation
- Video: Mechanical Weathering
- Reading Passage: Field Trip to Yosemite
- Video: Erosion by Water
- Video: Unique Properties of Water
- Video: Chemical Weathering

TEACHER NOTE Misconception: Students may think that most rocks are too hard to break into tiny pieces without using machines, or they may confuse the term mechanical with machine-driven processes. In fact, mechanical refers to physical processes (as opposed to chemical processes). Common natural processes, such as growing plant roots, blowing wind, and freezing and flowing of water, can break down rock mechanically. Students may see mechanical weathering referred to as physical weathering in some resources.

TEACHER NOTE Misconception: Students may think that only strong chemicals can break down rocks chemically. In fact, weak acids are also agents of chemical weathering. Even water can break down some rocks chemically.

TEACHER NOTE Practices: Science and Engineering Practice: Planning and Carrying Out Investigations: In this activity, students plan and carry out an investigation collaboratively. They make directional hypotheses about what happens to carbonate rocks as they manipulate the acidity of solutions with which the rocks come into contact. Then, they use the results of their investigation as evidence for chemical weathering. To extend this activity, ask students to develop a hypothesis about what would happen if a different independent variable in the experiment were manipulated, such as the size of the rock pieces or the type of rocks tested. If time permits, students can conduct their new experiments to test their hypothesis.

- Hands-On Lab: Chemical Weathering

Formative Assessment:
Throughout Explore, Technology Enhanced Items (TEIs) are embedded as multi-dimensional formative checks for understanding. You can use the data they provide to

- assign additional support
- extend learning
- design additional learning tasks to clarify student misconceptions

The Explore TEIs provide students with three attempts to demonstrate their proficiency. Scaffolded feedback is provided for each attempt. If a student does not achieve proficiency by the third attempt, a media asset is provided as an additional learning opportunity.

> **TEACHER NOTE Connections: Crosscutting Concept: Cause and Effect:** In these items, students suggest cause-and-effect relationships to explain and predict changes to Earth's surface caused by weathering processes. After students complete these items, you can use the second TEI as a starting point for a discussion of weathering in urban areas that might not be occurring in rural or wilderness areas. Students can work individually or in small groups to research a specific urban problem that leads to chemical weathering or a specific structure that is experiencing chemical weathering due to urban problems like pollution.

TEI Weathering by Water

TEI Water and Chemical Weathering

- Exploration: Mechanical and Chemical Weathering
- Image: Rate of Weathering
- Image: Differential Weathering

> **TEACHER NOTE Connections: Crosscutting Concept: Stability and Change:** In this item, students recognize that many variables contribute to the rate of change in a system and quantify changes in systems caused by mechanical and chemical weathering. To extend this activity, instruct students to work in small groups to create lists of two or three other changes that would affect the rate of weathering in Horseshoe Bend. Then, groups can exchange lists and determine whether weathering would happen more quickly or more slowly in each situation.

TEI Horseshoe Bend, Arizona

Explore More Resources

Resources in Explore More Resources support differentiation within your classroom by

- providing additional visualization of content
- affording extension of content to those students ready for acceleration
- offering Lexile reading levels for reading passages

Online explorations and hands-on experiences are provided so that students can conduct virtual investigations, collect and design investigations, and collect and analyze data; these skills are essential to developing scientific understanding.

Explain `45–90 minutes`

In Explore, students
1. uncovered scientific understandings
2. conducted investigations
3. analyzed data, text, and other media resources
4. collected evidence to support their scientific explanation

In Explain, provide students with time to formally compose their scientific explanations around the Explain or student-generated questions using evidence collected from Explore.

Scientific explanations are student responses, either written or orally presented, that explain scientific phenomena based upon evidence. Developing a scientific explanation requires students to analyze and interpret data to construct meaning out of the data. There are three main components to the scientific explanation: the claim, the evidence, and the reasoning.

To help students to communicate their scientific explanations, allow them to utilize the multimedia creation tools such as Board Builder and Whiteboard. Remind them that they may upload image, audio, and video files using the "attach file" option to communicate their scientific explanations.

Students may construct their scientific explanations individually or within a small group of students. Students should communicate their explanations with other classmates, and provide constructive criticism and refine their explanations prior to submission to the teacher. If explanations are used as a formative assessment, you can provide additional feedback and comments to support students as they refine their explanations.

EXPLAIN

What effect does weathering have on the surface of landforms such as mountains, buttes, mesas, and plateaus?

Elaborate with STEM   45–135 minutes

*Elaborate with STEM are optional extension resources available after students have demonstrated proficiency with standards addressed previously in the concept.

NGSS Components

SEP	CCC
■ Planning and Carrying Out Investigations ■ Constructing Explanations and Designing Solutions	■ Cause and Effect ■ Stability and Change

STEM In Action   45 minutes

STEM in Action ties the scientific concepts to real-world applications, with many connecting to STEM careers. Technology Enhanced Items (TEIs) expect students to critically read the Core Interactive Text (CIT) and review the provided media resources.

Applying Mechanical and Chemical Weathering

- Core Interactive Text: Applying Mechanical and Chemical Weathering
- Video: Cave Formation: The Role of Weathering and Erosion
- Image: The Sphinx
- Image: Acid Rain Damage to Marble Building Decoration
- Video: Weathering, Soil, and Erosion

TEACHER NOTE This summative assessment is designed to assess student understanding of the effects of physical and chemical weathering on man-made objects. Students may analyze the images as part of a think-pair-share, and then write their answers to the questions. Students can compare their answers as a whole class, if desired.

TEI **Weathering and the Moai**

STEM Project Starters

STEM Project Starters provide additional real-world contexts that require students to apply and extend their content knowledge related to the concept. STEM Project Starters can also serve as an alternative instructional hook presented at the beginning of the learning progression. The project can then be revisited throughout and at the end of the 5E learning cycle, for students to apply content knowledge.

STEM Project Starter: Weathering and Water

Recommended 90 minutes

What is the role of water in weathering?

> **TEACHER NOTE** This summative assessment combines science and mathematics to test students' understanding of weathering and requires them to think deeper about the effect weathering has on landforms. Through this activity students will work against common misconceptions such as the fact that mechanical weathering is caused through machine-driven processes and chemical weathering is caused only by strong chemicals. Students may complete this activity as homework or as part of a think-pair-share activity in class.

TEI Weathering and Water

STEM Project Starter: Saving Buildings

Recommended 60 minutes

What can scientists do to prevent the effects of chemical weathering, such as those caused by acid rain?

> **TEACHER NOTE** This STEM project relates to the process of protecting statues and buildings to prevent the effects of chemical weathering. It gives students a chance to research new technology and methods to help reduce the effects of chemical weathering on the world around them. Students must write a letter to a client explaining how they will protect a building.
>
> Guide students to structure their letters correctly. They should include a heading, the recipient's address, a greeting or salutation, the body of the letter, a complimentary close, and a signature line. The body of the letter should include an introduction, a paragraph describing the causes and effects of acid rain, a paragraph outlining plans to protect the building, and a conclusion. Students may complete this activity as a homework assignment or work in pairs to conduct research and come up with a common solution as part of an in-class activity. Pair English language learners with fluent English speakers to discuss ideas verbally before writing their letters. Have students work with a partner to review one another's writing and provide feedback. Students can use this feedback to improve their structure, transitions, clarity, and style.

TEI Saving Buildings

STEM Project Starter: Modeling Mechanical Weathering

Recommended 60 minutes

How can breaking pieces of candy model the effects of mechanical weathering?

TEACHER NOTE This summative assessment measures a student's understanding of mechanical weathering and addresses a common misconception that machine-driven processes cause mechanical weathering. Students incorporate principles of science and engineering as they design an experiment that simulates a natural process to simulate weathering, such as the blowing of wind or freezing or flowing of water.

To help students come up with an idea for an experiment, the teacher can provide a sample objective, materials list, explanation of safety precautions, and procedure for the hard chocolate-coated candies experiment described in the project description. Teacher may also demonstrate the experiment and discuss the results with students.

Students may design their experiments as pairs or in small groups. To extend the activity, the teacher may opt to pick one or two student-designed experiments to conduct in class.

TEI Modeling Mechanical Weathering

- Activity: Engineering Design Sheet

Evaluate 45–90 minutes

Explain Question:

What effect does weathering have on the surface of landforms such as mountains, buttes, mesas, and plateaus?

Lesson Questions (LQ):

- How do mechanical and chemical weathering change Earth's surface?

Throughout instruction and the 5E learning cycle, you will have collected formative assessment data to drive the assignment of resources and experiences to students. Evaluate is intended to include summative assessment checks for proficiency. You can use the Explain and Lesson Questions for the concept as a summative assessment in a variety of ways such as these:

- Post each Lesson Question (LQ) in various locations in the classroom, and have small groups of students generate claim statements related to the Lesson Question (LQ). Other students can add to the claim, or refute the claim, during a gallery walk where they place additional pieces of evidence on each Lesson Question (LQ) poster.
- Assign small groups of students to each Lesson Question (LQ) and have the groups generate a poster, board, graphic, or piece of text that answers the question. Use a jigsaw approach and create a second set of groups that contain members from each Lesson Question (LQ) group to share their ideas.
- Ask students to return to their initial ideas for the Explain question and add additional details and evidence.

Encourage students to review the concept review and complete the Student Self-Check practice assessment prior to assigning the Summative Teacher Concept assessment.

- Student Review and Practice Assessment
- Teacher Concept Assessments

Erosion and Deposition

The Five E Instructional Model

Science Techbook follows the 5E instructional model. This Model Lesson includes strategies for each of the 5Es. As you design the inquiry-based learning experience for students, be sure to collect data during instruction to drive your instructional decisions. Point-of-use teacher notes are also provided within each E-tab.

Engage   45–90 minutes

Engage Media Resources

The resources found in Engage are intended to stimulate students by exposing them to a phenomenon relevant to the content of the lesson. Engage also provides examples of relevant real-world applications that allow students to begin to make observations and relate the science content to their everyday lives. The Core Interactive Text (CIT) and media resources are carefully designed to prompt students to begin asking questions that they can investigate during the Explore phase of the lesson. They should also start collecting evidence to address the Explain question located at the bottom of the Engage page.

> **TEACHER NOTE** **Investigative Phenomenon:** Ask students whether they have ever traveled specifically to see breathtaking scenery. If so, what landforms did they find most spectacular? Suggest that these destinations were formed by eons of erosion and deposition. Focus on one such landform, the waterfall. Have students examine the picture of Yosemite Falls and the video about Niagara Forms. From what they have observed, are these erosional or depositional landforms? How does the process of erosion at Niagara Falls work? Yosemite Falls drops off solid granite; how do they think the process of erosion at Yosemite might differ from that of Niagara? Which falls is likely to be around the longest? Why? Have them consider: What happens to the rock eroded from these falls? Call on students to share their ideas. Have them capture a summary of the class's ideas using Studio.

- Core Interactive Text: Thinking about Erosion and Deposition
- Image: Yosemite Falls, California
- Video: Niagara Falls: A Force of Nature
- Video: Reshaping the Niagara Falls

Explain Question

The Explain question focuses students on gathering information in the Explore section. The Explain question can be used to

- Record what students already know related to the Explain question.
- Serve as a template or model for students to generate their own scientific questions.
- Collect evidence as students work through the lesson.
- Allow students to reflect on their growth before and after the lesson.

Explain

How do erosion and deposition shape Earth's surface?

- Image: The Power of Erosion

Engage Formative Assessment

Technology Enhanced Items (TEIs) found on the Engage page enable you to collect data on students' prior knowledge and identify the common misconceptions they may possess that are related to the topic of study. These items are designed as quick checks for understanding and allow each student one attempt at each question. You can use the data collected to decide whether to assign additional resources to the class, or determine what individual or groups of students may need reinforcement or accelerated learning, prior to completing the Explore portion of the lesson.

> **TEACHER NOTE** Use this formative assessment to evaluate students' prior knowledge of the concept. The Model Lesson provides information on common student misconceptions. Students often confuse erosion with weathering or deposition. In fact, weathering is the breaking down of rock into sediment by physical and/or chemical processes. (Technically, broken fragments of rock are called detritus, and sediment refers specifically to detritus that has been laid down by deposition; however, sediment is commonly used to describe all manner of rock fragments, and this is how the term is used in this lesson.)

TEI Differences

Before You Begin

What Do I Already Know about Erosion and Deposition?

> **TEACHER NOTE** This formative assessment is intended to provide the teacher with feedback on prior knowledge of this topic. Students should know that erosion is the removal of sediment by the action of gravity, water, ice, wind, or other agents. Deposition is the accumulation of eroded sediment in a new area due to the action of gravity, water, ice, or wind. Rocks are first affected by weathering; the resulting sediment is then affected by erosion, followed by deposition. Students should complete this activity individually.

TEI Agents of Erosion

> **TEACHER NOTE** This formative assessment gives the teacher information about student misconceptions. Students may believe that erosion always happens quickly as in a landslide or avalanche, while erosion is typically a very slow process that happens over long periods of time. Have students discuss the question as a class before entering their answers.

TEI Erosion through the Ages

TEACHER NOTE This formative assessment is intended to provide the teacher with feedback on prior knowledge of this topic. Its main focus is to determine whether students are aware humans can increase erosion rates. Students may believe that erosion always affects humans negatively. The teacher may find it valuable to discuss positive impacts of erosion and deposition, such as deposition of rich farming soils by glaciers and rivers.

TEI **Humans and Erosion**

TEACHER NOTE This formative assessment addresses possible misconceptions students have about the differences among erosion, weathering, and deposition. While students may believe erosion and weathering are synonymous, weathering describes breaking apart rock by physical or chemical means, whereas erosion describes moving matter from one place to another. Deposition, meanwhile, is the buildup of matter where it had not previously been.

TEI **What Type Is It?**

- Video: Rocks and Minerals
- Video: Phases of Matter
- Video: Heat and Matter

Explore   135 minutes

Lesson Questions (LQs):

1. How do rock and soil undergo erosion and deposition?
2. What are the specific agents of erosion, and what factors control deposition?
3. How can engineering design solve problems caused by erosion?

Effective science instruction involves a student-centered rather than a teacher-centered approach. This can be accomplished either with Directed Inquiry or Guided Inquiry, depending on the needs and abilities of your class. Encourage students to select a variety of resources in their pursuit of answers as they work through Explore, with the end goal of constructing their scientific explanation in the Explain tab.

Directed Inquiry	Guided Inquiry
In Directed Inquiry, teachers provide students with a sequence of specified resources, challenging questions, and clear outcomes. Within this context students are given the opportunity to interact independently with each resource as prescribed by the teacher. Often different students groups can be guided through several different resources at the same time. For example, one group could work on a reading passage while a second group conducts a small-group Hands-On Activity with the teacher, and a third group is independently engaged with an online interactive resource.	In Guided Inquiry, students have independence to decide the scope and sequence of their investigations. Using resources from Techbook, students determine for themselves which resources they will Explore to answer the Lesson Questions. It is important to note that each student will choose multiple resources, but no one student is expected to use all the resources available. Students also determine the order in which to explore these resources and how to record their findings.

NGSS Components

SEP	CCC
■ Developing and Using Models ■ Constructing Explanations and Designing Solutions	■ Cause and Effect ■ Structure and Function ■ Stability and Change

Lesson Question: How Do Rock and Soil Undergo Erosion and Deposition?

Recommended 30 minutes

TEACHER NOTE Crosscutting Concepts: Cause and Effect: In this concept, students suggest cause and effect relationships to explain and predict behaviors in Earth systems. They also propose causal relationships between Earth systems and landforms by examining what is known about smaller-scale mechanisms within the system, such as weathering and erosion. To introduce students to these relationships, use a Get VENN-y With It strategy with the images Erosion and Mitten Butte, Front View, which show small-scale and large-scale effects of erosion. This strategy is found on the Professional Learning tab. Click on Strategies & Resources, then Spotlight On Strategies (SOS). Get VENN-y With It is found underneath Cites Evidence.

- Core Interactive Text: How Do Rock and Soil Undergo Erosion and Deposition?
- Image: Erosion
- Image: Mitten Butte, Front View
- Reading Passage: Surviving Acid Rain
- Exploration: Erosion and Deposition
- Video: Tectonics and Surface Processes

Formative Assessment:

Throughout Explore, Technology Enhanced Items (TEIs) are embedded as multi-dimensional formative checks for understanding. You can use the data they provide to

- assign additional support
- extend learning
- design additional learning tasks to clarify student misconceptions

The Explore TEIs provide students with three attempts to demonstrate their proficiency. Scaffolded feedback is provided for each attempt. If a student does not achieve proficiency by the third attempt, a media asset is provided as an additional learning opportunity.

TEACHER NOTE Crosscutting Concepts: Stability and Change: In this item, students understand how much of geology deals with constructing explanations of how things change and how they remain stable. Students model changes in Earth's landforms over very long periods of time by ordering the natural processes that change Earth's surface. If necessary, review tectonic plate interactions with students before they complete this item. Write *convergent boundary, divergent boundary, transform boundary, fold mountains, fault-block mountains*, and other key terms on the board. Have students work in small groups to write definitions for the terms. Then, come together as a class and create class definitions from students' ideas.

TEI From Lakebed to Mountaintop

Lesson Question: What Are the Specific Agents of Erosion, and What Factors Control Deposition?

Recommended 50 minutes

- Core Interactive Text: What Are the Specific Agents of Erosion, and What Factors Control Deposition?
- Video: Erosion by Water
- Video: Wind Erosion
- Reading Passage: Landslide!
- Image: Moraine
- Video: Erosion and Deposition by Ice
- Image: Wentworth Scale
- Image: Hjulstrom's Diagram

TEACHER NOTE Science and Engineering Practices: Developing and Using Models: In this activity, students develop and use a model to generate data to support explanations of erosion and deposition, predict patterns of erosion and deposition, and analyze the factors that can influence the rate of erosion and deposition. After students complete the lab, if time permits, have students work in small groups to create a short presentation predicting how erosion and deposition occur with ice, wind, gravity, or lava.

- Hands-On Lab: Streaming Water

Lesson Question: How Can Engineering Design Solve Problems Caused by Erosion?

Recommended 55 minutes

TEACHER NOTE Misconception: Students may believe that erosion always affects Earth and humans negatively. In fact, erosion is simply the process of moving sediment from one location to another; it plays a crucial role in shaping Earth's surface over time. However, erosion—especially when caused or exacerbated by human activities such as deforestation—can remove nutrient-rich topsoil from an area and pollute water by increasing turbidity.

- Core Interactive Text: How Can Engineering Design Solve Problems Caused by Erosion?
- Image: Erosion

TEACHER NOTE Science and Engineering Practice: Constructing Explanations and Designing Solutions: In this activity, students design, evaluate, and refine a solution to a complex real-world problem of erosion that threatens a village. They base their solution on scientific knowledge, criteria, and tradeoff considerations. They construct and test a prototype of their solution and identify ways to improve their designs. To extend this activity, give students the opportunity to work in small groups to research erosion problems in their area and what is being done to address them, if anything. Students can present their findings as a presentation, report, or letter to an elected official calling attention to the problem or suggesting solutions.

- Hands-On Lab: Can You Save Sandy Village from Erosion?
- Image: Sand Dunes
- Image: Niagara Falls
- Video: Reshaping the Niagara Falls

TEACHER NOTE **Crosscutting Concepts: Structure and Function:** In this item, students investigate systems of erosion protection by examining the properties of different methods, the structures the methods use, and the interconnections between the properties and structures of the methods and the environment in which the erosion occurs. To extend this activity, give students time to work independently to find additional images of erosion, using Discovery Education resources or Internet sources. Students can then exchange images with a partner and discuss the system of erosion prevention that could be used in each situation, supporting their answers with evidence about how the properties and structures these methods use will work with the specific type of erosion.

TEI **How Can Engineers Protect Areas from Erosion?**

Explore More Resources

Resources in Explore More Resources support differentiation within your classroom by

- providing additional visualization of content
- affording extension of content to those students ready for acceleration
- offering Lexile reading levels for reading passages

Online explorations and hands-on experiences are provided so that students can conduct virtual investigations, collect and design investigations, and collect and analyze data; these skills are essential to developing scientific understanding.

Explain `45–90 minutes`

In Explore students
1. uncovered scientific understandings
2. conducted investigations
3. analyzed data, text, and other media resources
4. collected evidence to support their scientific explanation

In Explain, provide students with time to formally compose their scientific explanations around the Explain or student-generated questions using evidence collected from Explore.

Scientific explanations are student responses, either written or orally presented, that explain scientific phenomena based upon evidence. Developing a scientific explanation requires students to analyze and interpret data to construct meaning out of the data. There are three main components to the scientific explanation: the claim, the evidence, and the reasoning.

To help students to communicate their scientific explanations, allow them to utilize the multimedia creation tools such as Board Builder and Whiteboard. Remind them that they may upload image, audio, and video files using the "attach file" option to communicate their scientific explanations.

Students may construct their scientific explanations individually or within a small group of students. Students should communicate their explanations with other classmates, and provide constructive criticism and refine their explanations prior to submission to the teacher. If explanations are used as a formative assessment, you can provide additional feedback and comments to support students as they refine their explanations.

EXPLAIN

How do erosion and deposition shape Earth's surface?

Elaborate with STEM `45–90 minutes`

*Elaborate with STEM are optional extension resources available after students have demonstrated proficiency with standards addressed previously in the concept.

NGSS Components

SEP	CCC
■ Asking Questions and Defining Problems ■ Planning and Carrying Out Investigation ■ Using Mathematics and Computational Thinking ■ Constructing Explanations and Designing Solutions ■ Obtaining, Evaluating, and Communicating Information	■ Cause and Effect ■ Structure and Function

STEM In Action `45 minutes`

STEM in Action ties the scientific concepts to real-world applications, with many connecting to STEM careers. Technology Enhanced Items (TEIs) expect students to critically read the Core Interactive Text (CIT) and review the provided media resources.

Applying Erosion and Deposition

- Core Interactive Text: Applying Erosion and Deposition
- Video: Protecting the Coasts
- Image: Erosion
- Video: Erosion Threatens National Treasures
- Video: Value in Coastline

TEACHER NOTE This assessment incorporates math into what students have learned about erosion. Students will have to use an equation to determine the rate of erosion of a beach.

TEI Calculating Rate of Erosion

STEM Project Starters

STEM Project Starters provide additional real-world contexts that require students to apply and extend their content knowledge related to the concept. STEM Project Starters can also serve as an alternative instructional hook presented at the beginning of the learning progression. The project can then be revisited throughout and at the end of the 5E learning cycle, for students to apply content knowledge.

STEM Project Starter: Can You Save Sandy Village from Erosion?

`Recommended 45 minutes`

How can you protect both humans and the environment when preventing erosion?

> **TEACHER NOTE** This STEM project builds on students' experience in the Hands-On Lab Can You Save Sandy Village from Erosion? It asks students to connect the lab to real-life situations and consider solutions to problems faced by engineers, developers, and conservationists.

TEI Lessons from Sandy Village

- Activity: Engineering Design Sheet

STEM Project Starter Erosion and Deposition

`Recommended 45 minutes`

Can you explore erosion and deposition through a glacier?

> **TEACHER NOTE** The exploration includes technology, science, and engineering. The assessment builds on students' knowledge by encouraging them to consider the use of the device they have digitally explored.

- Exploration: Erosion and Deposition

TEI Using Core Samples

Evaluate `45–90 minutes`

Explain Question:
How do erosion and deposition shape Earth's surface?

Lesson Questions (LQ):
- How do rock and soil undergo erosion and deposition?
- What are the specific agents of erosion, and what factors control deposition?
- How can engineering design solve problems caused by erosion?

Throughout instruction and the 5E learning cycle, you will have collected formative assessment data to drive the assignment of resources and experiences to students. Evaluate is intended to include summative assessment checks for proficiency. You can use the Explain and Lesson Questions for the concept as a summative assessment in a variety of ways such as these:

- Post each Lesson Question (LQ) in various locations in the classroom, and have small groups of students generate claim statements related to the Lesson Question (LQ). Other students can add to the claim, or refute the claim, during a gallery walk where they place additional pieces of evidence on each Lesson Question (LQ) poster.
- Assign small groups of students to each Lesson Question (LQ) and have the groups generate a poster, board, graphic, or piece of text that answers the question. Use a jigsaw approach and create a second set of groups that contain members from each Lesson Question (LQ) group to share their ideas.
- Ask students to return to their initial ideas for the Explain question and add additional details and evidence.

Encourage students to review the concept review and complete the Student Self-Check practice assessment prior to assigning the Summative Teacher Concept assessment.

- Student Review and Practice Assessment
- Teacher Concept Assessments

The History of Life on Earth

The Five E Instructional Model

Science Techbook follows the 5E instructional model. This Model Lesson includes strategies for each of the 5Es. As you design the inquiry-based learning experience for students, be sure to collect data during instruction to drive your instructional decisions. Point-of-use teacher notes are also provided within each E-tab.

Engage 45–90 minutes

Engage Media Resources

The resources found in Engage are intended to stimulate students by exposing them to a phenomenon relevant to the content of the lesson. Engage also provides examples of relevant real-world applications that allow students to begin to make observations and relate the science content to their everyday lives. The Core Interactive Text (CIT) and media resources are carefully designed to prompt students to begin asking questions that they can investigate during the Explore phase of the lesson. They should also start collecting evidence to address the Explain question located at the bottom of the Engage page.

> **TEACHER NOTE** **Investigative Phenomenon:** Since the discovery of *Archaeopteryx* in Bavaria in 1860, scientists have been intrigued by the evolutionary relationship between reptiles and birds. Recent discoveries, particularly in China, provide evidence of a close relationship between dinosaurs and modern birds. Use the text and video "Living Dinosaurs" to show how, as our knowledge of the fossil record improves, so does our understanding of the relationship between past animal groups and those of modern day. The existence of "missing links," or big gaps in the fossil record, is often used as an argument against evolution. Have students discuss how the discovery of fossils in China has closed the gap for birds. As they continue through Engage, ask them whether new finds of human-like fossils are closing the gaps in our understanding of the evolution of humans. How will the discovery of more fossils provide additional evidence for organisms that evolved from a common ancestor?

- Core Interactive Text: Examining the History of Life on Earth
- Video: Living Dinosaurs
- Image: Lucy, *Australopithecus afarenis*
- Video: Evidence for Evolution

Explain Question

The Explain question focuses students on gathering information in the Explore section. The Explain question can be used to

- Record what students already know related to the Explain question.
- Serve as a template or model for students to generate their own scientific questions.
- Collect evidence as students work through the lesson.
- Allow students to reflect on their growth before and after the lesson.

Explain
How can scientists construct a timeline for the history of life on Earth?

■ Image: Brown Mouse

TEACHER NOTE Use this formative student response to evaluate students' prior knowledge of the concept. Student responses will vary with their understanding of the topic.

TEI Mice and Humans

Engage Formative Assessment
Technology Enhanced Items (TEIs) found on the Engage page enable you to collect data on students' prior knowledge and identify the common misconceptions they may possess that are related to the topic of study. These items are designed as quick checks for understanding and allow each student one attempt at each question. You can use the data collected to decide whether to assign additional resources to the class, or determine what individual or groups of students may need reinforcement or accelerated learning, prior to completing the Explore portion of the lesson.

Before You Begin
What Do I Already Know about the History of Life on Earth?

TEACHER NOTE Use this formative assessment to gauge student misconceptions that the highest organism on an evolutionary diagram is the most advanced or specialized. Cladograms such as this one represent only the relationships between organisms, not their degree of specialization. Students may discuss the question with a partner before entering their answers.

TEI Reading a Cladogram

TEACHER NOTE This formative assessment gives the teacher understanding of students' prior knowledge of classification of organisms. Students do not need to be able to trace an entire taxonomic classification from memory, but should have a basic understanding of the major differences between types of organisms. Students should answer this question individually.

TEI Understanding Taxonomy

TEACHER NOTE This formative assessment probes students' understanding and prior knowledge of evolution and addresses common misconceptions. Students may think that evolutionary theory describes the origin of life, whereas it addresses the relationship between species. They may think that one species must evolve into only one other species, whereas a single species can give rise to several other species (adaptive radiation). Students may think that humans and chimpanzees are distantly related, whereas chimpanzees are in fact our closest extant relatives. Students should answer the question individually.

TEI Evolutionary Theory

TEACHER NOTE This formative assessment gives the teacher information about student misconceptions related to extinction and biodiversity. Students who answer true for the first statement may think that biodiversity refers to the number of individuals in a habitat (for example, a large herd of wild horses), when actually the number of species determines biodiversity. Students who answer true for the second statement may think that extinction is uncommon and only follows disaster. Extinction actually occurs all the time, usually following small-scale events. Students who answer false to the last question may think that humans lived during the age of the dinosaurs. In fact, dinosaurs mostly died out 65 million years ago, while humans did not arise until between 10 and 2 million years ago. You may wish to lead a class discussion about student responses after students have entered their answers individually.

TEI Evolution and Biodiversity

- Video: Living Things
- Video: Classifying Living Things
- Video: Ecosystems and Adaptations
- Video: The Evolution of Complex Organs

Explore `180 minutes`

Lesson Questions (LQs):

1. What are different scientific explanations for how and when life on Earth evolved?
2. What are biological diversity, episodic speciation, and mass extinction?
3. What is an evolutionary tree diagram?
4. What are some possible scientific explanations for aspects of the fossil record such as gaps and the sequential nature of fossils?

Effective science instruction involves a student-centered rather than a teacher-centered approach. This can be accomplished either with Directed Inquiry or Guided Inquiry, depending on the needs and abilities of your class. Encourage students to select a variety of resources in their pursuit of answers as they work through Explore, with the end goal of constructing their scientific explanation in the Explain tab.

Directed Inquiry	Guided Inquiry
In Directed Inquiry, teachers provide students with a sequence of specified resources, challenging questions, and clear outcomes. Within this context students are given the opportunity to interact independently with each resource as prescribed by the teacher. Often different students groups can be guided through several different resources at the same time. For example, one group could work on a reading passage while a second group conducts a small-group Hands-On Activity with the teacher, and a third group is independently engaged with an online interactive resource.	In Guided Inquiry, students have independence to decide the scope and sequence of their investigations. Using resources from Techbook, students determine for themselves which resources they will Explore to answer the Lesson Questions. It is important to note that each student will choose multiple resources, but no one student is expected to use all the resources available. Students also determine the order in which to explore these resources and how to record their findings.

NGSS Components

SEP	CCC
■ Developing and Using Models ■ Constructing Explanations and Designing Solutions ■ Engaging in Argument from Evidence	■ Cause and Effect

Lesson Question: What Are Different Scientific Explanations for How and When Life on Earth Evolved?

`Recommended 45 minutes`

TEACHER NOTE **Practices: Science and Engineering Practice: Engaging in Argument from Evidence:** In this lesson, students will learn about evolution and the beginnings of life, both topics that get a lot of attention and debate. It is, therefore, important that students are given the opportunity to evaluate the claims, evidence, and reasoning behind these ideas, which are currently accepted explanations for the way life on Earth began and has evolved. In this way, students can determine the merits of the arguments in an educated manner. Provide students with opportunities to debate the arguments that support the different ideas behind how life came to be on Earth and how organisms evolved. For instance, break the students up into groups that support particular mindsets regarding evolution. The students do not necessarily have to support their assigned viewpoint, but they should be able to discuss these topics from all sides (indeed, they should support the viewpoints that have scientific evidence). In discussing different viewpoints, students will better understand the ideas supported by scientific evidence.

- Core Interactive Text: What Are Different Scientific Explanations for How and When Life on Earth Evolved?
- Video: The Formation of Earth
- Video: Evolution of the Biosphere
- Video: From the Big Bang to Early Life
- Video: The Evolution of Complex Organisms
- Video: Photosynthesis and Early Life
- Image: Trilobite Fossil
- Video: Ardipithecus and the Modern World

Formative Assessment

Throughout Explore, Technology Enhanced Items (TEIs) are embedded as multi-dimensional formative checks for understanding. You can use the data they provide to

- assign additional support
- extend learning
- design additional learning tasks to clarify student misconceptions

The Explore TEIs provide students with three attempts to demonstrate their proficiency. Scaffolded feedback is provided for each attempt. If a student does not achieve proficiency by the third attempt, a media asset is provided as an additional learning opportunity.

TEACHER NOTE Practices: Science and Engineering Practice: Constructing Explanations and Designing Solutions: In this item, students will apply their understanding of the theory of evolution and the model shown to explain the extent to which the fossil record supports and provides evidence for the theory of evolution. Students will construct an explanation based on evidence and the assumption that the theory of evolution describes the natural world around them. To extend this item, invite students to work in groups of two. Provide groups with pictures of fossils of other organisms (for example, a bird or whale). Ask students to conduct research on the organism and place the fossils on a timeline to show the progression of the group over time. In a whole-class discussion, students will explain how the changes seen in the fossil record have helped the organisms adapt to their environment.

TEI **TEI: Fossils**

Lesson Question: What Are Biological Diversity, Episodic Speciation, and Mass Extinction?

Recommended 30 minutes

- Core Interactive Text: What Are Biological Diversity, Episodic Speciation, and Mass Extinction?
- Video: The Tropics
- Video: Biodiversity
- Video: Biological Diversity, Episodic Speciation, and Mass Extinction
- Video: Mass Extinction and Adaptive Radiation

TEACHER NOTE Connections: Crosscutting Concept: Cause and Effect: In this item, students will suggest and predict cause-and-effect relationships for complex natural systems by examining what is known about the mechanisms within the system. To extend this item, place students in groups of three or four. Provide each group with examples of human-driven and natural causes of environmental change. Examples may include pollution and volcanic eruptions. Instruct students to determine how each change will affect the organisms in the area. Students should explain how they think different species will respond to the change and also cite evidence that may support their ideas, such as fossil evidence.

TEI **Changes in the Environment**

Lesson Question: What Is an Evolutionary Tree Diagram?

Recommended 45 minutes

- Core Interactive Text: What Is an Evolutionary Tree Diagram?
- Exploration: The History of Life on Earth
- Video: Evolutionary Trees and Classification
- Video: Evolutionary Tree and Sequence of Life on Earth

TEACHER NOTE Practices: Science and Engineering Practice: Developing and Using Models: In this item, students use a model of a phylogenic tree to illustrate and predict relationships among the components of the model in their natural setting. They synthesize information from the model to show the relationships among variables as they relate to classification within the system of living things. To extend this item, instruct students to work in groups of two to develop a phylogenic tree of an animal of their choosing. Their phylogenic tree should include identifying characteristics that unify the different groups within the evolutionary tree they create. Limit the number of "branches" on the tree as appropriate.

TEI Interpreting a Phylogenic Tree

Lesson Question: What Are Some Possible Scientific Explanations for Aspects of the Fossil Record Such as Gaps and the Sequential Nature of Fossils?

Recommended 60 minutes

- Core Interactive Text: What Are Some Possible Scientific Explanations for Aspects of the Fossil Record Such as Gaps and the Sequential Nature of Fossils?
- Video: Interpreting the Fossil Record
- Video: How Complete Is the Fossil Record?
- Video: Ancient Life and the Fossil Record
- Image: The Evolution of the Whale
- Image: Evolution of the Foraminifera
- Video: Stratigraphic Principles

TEACHER NOTE: Connections: Crosscutting Concept: Cause and Effect: In this item, students suggest cause-and-effect relationships to explain and predict whether organisms are likely to fossilize. They recognize that these cause-and-effect relationships have an impact on the fossil record and the gaps that exist within it. To extend this item, students can work in groups of two to classify a variety of living things based on whether they are likely to fossilize. Students should include evidence to support their reasoning.

TEI Will They Fossilize?

Explore More Resources

Resources in Explore More Resources support differentiation within your classroom by

- providing additional visualization of content
- affording extension of content to those students ready for acceleration
- offering Lexile reading levels for reading passages

Online explorations and hands-on experiences are provided so that students can conduct virtual investigations, collect and design investigations, and collect and analyze data; these skills are essential to developing scientific understanding.

Explain `45–135 minutes`

In Explore, students
1. uncovered scientific understandings
2. conducted investigations
3. analyzed data, text, and other media resources
4. collected evidence to support their scientific explanation

In Explain, provide students with time to formally compose their scientific explanations around the Explain or student-generated questions using evidence collected from Explore.

Scientific explanations are student responses, either written or orally presented, that explain scientific phenomena based upon evidence. Developing a scientific explanation requires students to analyze and interpret data to construct meaning out of the data. There are three main components to the scientific explanation: the claim, the evidence, and the reasoning.

To help students to communicate their scientific explanations, allow them to utilize the multimedia creation tools such as Board Builder and Whiteboard. Remind them that they may upload image, audio, and video files using the "attach file" option to communicate their scientific explanations.

Students may construct their scientific explanations individually or within a small group of students. Students should communicate their explanations with other classmates, and provide constructive criticism and refine their explanations prior to submission to the teacher. If explanations are used as a formative assessment, you can provide additional feedback and comments to support students as they refine their explanations.

EXPLAIN

How can scientists construct a timeline for the history of life on Earth?

Elaborate with STEM `45–135 minutes`

*Elaborate with STEM are optional extension resources available after students have demonstrated proficiency with standards addressed previously in the concept.

NGSS Components

SEP	CCC
■ Asking Questions and Defining Problems ■ Developing and Using Models ■ Planning and Carrying Out Investigations ■ Analyzing and Interpreting Data ■ Using Mathematics and Computational Thinking ■ Constructing Explanations and Designing Solutions ■ Obtaining, Evaluating, and Communicating Information	■ Patterns ■ Cause and Effect ■ Scale, Proportion, and Quantity ■ Stability and Change

STEM in Action `45 minutes`

STEM in Action ties the scientific concepts to real-world applications, with many connecting to STEM careers. Technology Enhanced Items (TEIs) expect students to critically read the Core Interactive Text (CIT) and review the provided media resources.

Applying Explanations for the History of Life on Earth

TEACHER NOTE You may wish to lead a class discussion about the competing ideas regarding how life originated on Earth. Have students list the evidence for and against each idea.

- Core Interactive Text: Applying Explanations for the History of Life on Earth
- Video: Underwater Volcanic Nurseries
- Image: Black Smoker

TEACHER NOTE You may wish to have students check PubMed themselves to see that anyone can access information about the human genome. Discuss how this knowledge can be used to trace the history of life (for example, through comparing the relatedness of close and distant species).

- Video: The Human Genome Project

TEACHER NOTE Students complete the Hands-On Laboratory: Analyzing Biological Diversity, Extinction, and Evolution.

- Hands-On Laboratory: Analyzing Biological Diversity, Extinction, and Evolution

STEM Project Starters

STEM Project Starters provide additional real-world contexts that require students to apply and extend their content knowledge related to the concept. STEM Project Starters can also serve as an alternative instructional hook presented at the beginning of the learning progression. The project can then be revisited throughout and at the end of the 5E learning cycle, for students to apply content knowledge.

STEM Project Starter: A Map of Life `Recommended 60 minutes`

Can you map the entire history of life on Earth?

> **TEACHER NOTE** This summative project can be done immediately following the Hands-On Activity or at the end of the Elaborate section. Provide students with a larger or different map than they used in the Hands-On Activity. (For the greatest challenge, supply a map that can show a timeline going back 4.5 billion years.) Students should work in groups, which can be the same or different groups as they were in for the Hands-On Activity.

- Hands-On Activity: Life on Earth: A Timeline

`TEI` **Technologies**

STEM Project Starter: Saving Endangered Species `Recommended 90 minutes`

This endangered hummingbird is being banded for tracking purposes. How might you track an endangered species?

> **TEACHER NOTE** This summative project builds on student knowledge from the Hands-On Lab and extends the project to include an engineering component. Students should complete the project in small groups.

- Hands-On Lab: Analyzing Biological Diversity, Extinction, and Evolution

`TEI` **Attach Design**

- Activity: Engineering Design Sheet

STEM Project Starter: Fossil Heads `Recommended 45 minutes`

How can you use two-way tables to analyze morphological traits?

- Image: Idealized Morphology of Trilobites and Helemtiids
- Image: Non-Oblate Head Shield
- Video: DE Academy: Two-Way Table

`TEI` **Saber-Toothed Cats**

Evaluate `45–90 minutes`

Explain Question:
How can scientists construct a timeline for the history of life on Earth?

Lesson Questions (LQ):
- What are different scientific explanations for how and when life on Earth evolved?
- What are biological diversity, episodic speciation, and mass extinction?
- What is an evolutionary tree diagram?
- What are some possible scientific explanations for aspects of the fossil record such as gaps and the sequential nature of fossils?

Throughout instruction and the 5E learning cycle, you will have collected formative assessment data to drive the assignment of resources and experiences to students. Evaluate is intended to include summative assessment checks for proficiency. You can use the Explain and Lesson Questions for the concept as a summative assessment in a variety of ways such as these:

- Post each Lesson Question (LQ) in various locations in the classroom, and have small groups of students generate claim statements related to the Lesson Question (LQ). Other students can add to the claim, or refute the claim, during a gallery walk where they place additional pieces of evidence on each Lesson Question (LQ) poster.
- Assign small groups of students to each Lesson Question (LQ) and have the groups generate a poster, board, graphic, or piece of text that answers the question. Use a jigsaw approach and create a second set of groups that contain members from each Lesson Question (LQ) group to share their ideas.
- Ask students to return to their initial ideas for the Explain question and add additional details and evidence.

Encourage students to review the concept review and complete the Student Self-Check practice assessment prior to assigning the Summative Teacher Concept assessment.

- Student Review and Practice Assessment
- Teacher Concept Assessments

The Development of Earth

The Five E Instructional Model

Science Techbook follows the 5E instructional model. This Model Lesson includes strategies for each of the 5Es. As you design the inquiry-based learning experience for students, be sure to collect data during instruction to drive your instructional decisions. Point-of-use teacher notes are also provided within each E-tab.

Engage 45–90 minutes

Engage Media Resources

The resources found in Engage are intended to stimulate students by exposing them to a phenomenon relevant to the content of the lesson. Engage also provides examples of relevant real-world applications that allow students to begin to make observations and relate the science content to their everyday lives. The Core Interactive Text (CIT) and media resources are carefully designed to prompt students to begin asking questions that they can investigate during the Explore phase of the lesson. They should also start collecting evidence to address the Explain question located at the bottom of the Engage page.

> **TEACHER NOTE** **Investigative Phenomenon:** The nature of the big bang and the origins of our universe are difficult to understand. Start this concept by asking students how they envisage the universe came about (Some may consider it has always been in existence; others may provide a variety of creation myths.). Ask them why it so difficult for us to study the early universe and what evidence they consider exists for the various explanations available. Use the text and video "Everything Starts with the Big Bang" to stimulate discussion and supplement their understanding about exactly what the big bang was.

- Core Interactive Text: Reflecting on the Development of Earth
- Video: Everything Starts with the Big Bang
- Image: An Active Volcano
- Video: How Are Stars Formed?

Explain Question

The Explain question focuses students on gathering information in the Explore section. The Explain question can be used to

- Record what students already know related to the Explain question.
- Serve as a template or model for students to generate their own scientific questions.
- Collect evidence as students work through the lesson.
- Allow students to reflect on their growth before and after the lesson.

Explain

How can we explain the processes that led to the early conditions on Earth?

- Image: Volcanic Eruption

Engage Formative Assessment

Technology Enhanced Items (TEIs) found on the Engage page enable you to collect data on students' prior knowledge and identify the common misconceptions they may possess that are related to the topic of study. These items are designed as quick checks for understanding and allow each student one attempt at each question. You can use the data collected to decide whether to assign additional resources to the class, or determine what individual or groups of students may need reinforcement or accelerated learning, prior to completing the Explore portion of the lesson.

> **TEACHER NOTE** Use this formative assessment to evaluate prior knowledge of the concept. Students should understand that the Earth we see from the surface actually originated in the interior.

TEI Explain

Before You Begin

What Do I Already Know about The Development of Earth?

> **TEACHER NOTE** This two-part formative assessment is intended to provide the teacher with feedback on prior knowledge of the water and carbon cycles, which are important to understanding the evolution of Earth's atmosphere.
>
> Students may be under the misconception that the atmosphere surrounding Earth has always had same composition of gases it has today and that it has remained stable. In reality, the atmosphere has evolved and changed as Earth itself has evolved. Slow cooling of Earth has changed the composition of the gases that it releases. In addition, as we burn more fossil fuels, we are changing the concentration of carbon dioxide in the atmosphere.
>
> Allowing students to complete this individually will give you the best assessment; however, having them complete with a partner may generate discussion which activates their prior knowledge on the topic.

TEI Cycles

TEI False Statements

TEACHER NOTE This formative assessment will help you determine what students understand about current processes that form the geosphere. It will also advise you as to which students may hold the misconception that Earth is unchanging and once formed, remains uniform. The reality is that Earth is constantly reshaping its surface and is in still in the process of cooling from its original formation.

Allowing students to complete this individually will provide the best assessment; however, having them complete with a partner may generate discussions which activate their prior knowledge on the topic.

TEI Geosphere Creation

TEACHER NOTE This formative activity is intended to provide the teacher with feedback on prior knowledge of photosynthesis, which is important to understanding the history of Earth's atmosphere. Students must have a working knowledge of this process to understand how the atmosphere on Earth could have accumulated enough oxygen to support the life we see today. Students should have learned the basics of photosynthesis in middle school.

Allowing students to complete this individually will give you the best assessment; however, having them complete with a partner may generate discussion which activates their prior knowledge on the topic.

TEI Photosynthesis

TEI Summarize

- Video: Periodic Table: Periods and Groups
- Image: Time Scale
- Video: The Geologic Time Scale
- Video: Density

Explore — 90 minutes

Lesson Questions (LQs):

1. What were the conditions of early Earth?

Effective science instruction involves a student-centered rather than a teacher-centered approach. This can be accomplished either with Directed Inquiry or Guided Inquiry, depending on the needs and abilities of your class. Encourage students to select a variety of resources in their pursuit of answers as they work through Explore, with the end goal of constructing their scientific explanation in the Explain tab.

Directed Inquiry	Guided Inquiry
In Directed Inquiry, teachers provide students with a sequence of specified resources, challenging questions, and clear outcomes. Within this context students are given the opportunity to interact independently with each resource as prescribed by the teacher. Often different students groups can be guided through several different resources at the same time. For example, one group could work on a reading passage while a second group conducts a small-group Hands-On Activity with the teacher, and a third group is independently engaged with an online interactive resource.	In Guided Inquiry, students have independence to decide the scope and sequence of their investigations. Using resources from Techbook, students determine for themselves which resources they will Explore to answer the Lesson Questions. It is important to note that each student will choose multiple resources, but no one student is expected to use all the resources available. Students also determine the order in which to explore these resources and how to record their findings.

NGSS Components

SEP	CCC
■ Developing and Using Models ■ Planning and Carrying Out Investigations	■ Systems and System Models ■ Stability and Change

Lesson Question: What Were the Conditions of Early Earth?   **Recommended 90 minutes**

> **TEACHER NOTE** Practices: Science and Engineering Practice: Developing and Using Models: In this lesson, have students develop and revise models based on evidence to illustrate or predict the relationships of how the early characteristics and mechanisms of the Earth system have led to its current system. As a class, have students develop a model describing the present internal structure, atmosphere, and surface of Earth. Then, provide students with a simplified model of early Earth and ask them to discuss the relationship between the two models and predict the processes that transformed one system into the other. Return to this discussion as other models are encountered.

- Core Interactive Text: What Were the Conditions of Early Earth?
- Video: The Formation of Earth
- Image: Earth's Internal Layers

TEACHER NOTE Practices: Science and Engineering Practice: Planning and Carrying Out Investigations: In this Hands-On Lab activity, students will plan and test a design of a physical model of the internal structures of Earth using colored liquid layers to produce data that will serve as evidence supporting explanations for phenomena. It includes both direct inquiry about physical properties such as density and guided inquiries to predict the outcomes of the experiment. In this activity, students are required to plan the various investigation steps and determine the suitable composition (density) of the water solutions. Challenge the student to think of how this model represents Earth processes, and its limitations for doing so, and have each group record an original idea on the board for a class discussion.

- Hands-On Lab: Earth's Interior
- Reading Passage: The Iron Catastrophe
- Video: Earth's Internal Structure
- Video: The Formation of Earth's Moon

Formative Assessment:

Throughout Explore, Technology Enhanced Items (TEIs) are embedded as multi-dimensional formative checks for understanding. You can use the data they provide to

- assign additional support
- extend learning
- design additional learning tasks to clarify student misconceptions

The Explore TEIs provide students with three attempts to demonstrate their proficiency. Scaffolded feedback is provided for each attempt. If a student does not achieve proficiency by the third attempt, a media asset is provided as an additional learning opportunity.

TEACHER NOTE Practices: Science and Engineering Practice: Developing and Using Models: In this formative assessment, students are required to develop and use a model to illustrate the relationship between components of Earth's internal systems. This activity will help the students overcome the misconception that the Earth's interior is uniform. To extend this activity, separate the students into small groups and assign to each group a different parameter (temperature, pressure, composition, plasticity). Have each group discuss how that parameter allows description and classification of the various internal layers. Have groups summarize their findings and share with the class.

TEI Earth's Layers

- Image: Volcanic Gases
- Reading Passage: Chemistry of Volcanic Gases
- Video: Water on the Developing Earth

TEACHER NOTE **Connections: Crosscutting Concept: Systems and System Models:** In this item, students explore the boundaries and initial conditions of the three historical atmospheric systems. They will use a graphical model that can be used to predict the behavior of the system but has limited precision and reliability due to approximations inherent in models. To extend student thinking about the limitations and benefits of this model, ask: Why might the early hydrogen and helium atmosphere of the nascent Earth not have been included on this chart? How do the format and scale of this chart affect the precision of data provided about constituent gases? How does this visual representation expand our understanding of atmospheric changes?

TEACHER NOTE **Misconception:** Students may think that the composition of the atmosphere has not changed. In fact, Earth's atmosphere has evolved slowly over time, much like the surface of the planet. Early Earth had no atmosphere; the current composition (primarily nitrogen and oxygen with various trace gases) evolved over billions of years.

TEI **The Changing Atmosphere**

- Image: A New Ocean
- Video: Photosynthesis and Earth's Atmosphere
- Image: Living Stromatolites
- Image: Fossilized Stromatolite
- Video: Early Earth and Early Life
- Reading Passage: The Cambrian Explosion

TEACHER NOTE **Connections: Crosscutting Concept: Stability and Change:** In this formative assessment, students construct a valid explanation of changes in the Earth system, including its internal structure, atmosphere, and oceans, by examining the changes over time and forces acting on the Earth. Students should also be aware that the constituent elements have remained stable, despite these changes. To extend this lesson, ask students what kinds of samples they would look for to find evidence of the various phases of Earth's formation. Allow them to discuss their answers in small groups and justify their choices.

TEI **Earth's History**

Explore More Resources

Resources in Explore More Resources support differentiation within your classroom by

- providing additional visualization of content
- affording extension of content to those students ready for acceleration
- offering Lexile reading levels for reading passages

Online explorations and hands-on experiences are provided so that students can conduct virtual investigations, collect and design investigations, and collect and analyze data; these skills are essential to developing scientific understanding.

Explain `45–90 minutes`

In Explore, students
1. uncovered scientific understandings
2. conducted investigations
3. analyzed data, text, and other media resources
4. collected evidence to support their scientific explanation

In Explain, provide students with time to formally compose their scientific explanations around the Explainor student-generated questions using evidence collected from Explore.

Scientific explanations are student responses, either written or orally presented, that explain scientific phenomena based upon evidence. Developing a scientific explanation requires students to analyze and interpret data to construct meaning out of the data. There are three main components to the scientific explanation: the claim, the evidence, and the reasoning.

To help students to communicate their scientific explanations, allow them to utilize the multimedia creation tools such as Board Builder and Whiteboard. Remind them that they may upload image, audio, and video files using the "attach file" option to communicate their scientific explanations.

Students may construct their scientific explanations individually or within a small group of students. Students should communicate their explanations with other classmates, and provide constructive criticism and refine their explanations prior to submission to the teacher. If explanations are used as a formative assessment, you can provide additional feedback and comments to support students as they refine their explanations.

EXPLAIN?

How can we explain the processes that led to the early conditions on Earth?

Elaborate with STEM `45–135 minutes`

*Elaborate with STEM are optional extension resources available after students have demonstrated proficiency with standards addressed previously in the concept.

NGSS Components

SEP	CCC
■ Developing and Using Models ■ Planning and Carrying Out Investigations ■ Analyzing and Interpreting Data ■ Using Mathematics and Computational Thinking ■ Constructing Explanations and Designing Solutions ■ Engaging in Argument from Evidence ■ Obtaining, Evaluating, and Communicating Information	■ Cause and Effect ■ Scale, Proportion, and Quantity ■ Systems and System Models ■ Energy and Matter: Flows, Cycles, and Conservation ■ Structure and Function ■ Stability and Change

STEM In Action `45 minutes`

STEM in Action ties the scientific concepts to real-world applications, with many connecting to STEM careers. Technology Enhanced Items (TEIs) expect students to critically read the Core Interactive Text (CIT) and review the provided media resources.

Applying the Development of Earth

- Core Interactive Text: Applying the Development of Earth
- Video: Global Warming
- Image: Retreating Glacier
- Video: Radiometric Dating
- Image: Jack Hills, Australia
- Video: Geological Time

TEACHER NOTE This summative activity assesses how students are able to apply the information about STEM technology. This can be completed by individual students, or students may work with a partner to help them discuss the reasoning for their choices.

- Exploration: Absolute Dating

TEI Radiometric Dating

STEM Project Starters

STEM Project Starters provide additional real-world contexts that require students to apply and extend their content knowledge related to the concept. STEM Project Starters can also serve as an alternative instructional hook presented at the beginning of the learning progression. The project can then be revisited throughout and at the end of the 5E learning cycle, for students to apply content knowledge.

STEM Project Starter: Oldest Earth Samples `Recommended 45 minutes`

What technologies are used to determine the ages of Earth materials?

> **TEACHER NOTE** This summative assessment will allow you to determine if students understand the technology and science behind how scientists have dated Earth and how they determine conditions that existed in those early beginnings. Through their research, they will learn more about radioactive dating.
>
> Students may work independently; however, this may work well as a small group project, particularly groups of three, with each member researching one of the sample types.

TEI Oldest Earth Samples

STEM Project Starter Modeling Earth's Layers `Recommended 90 minutes`

What do Earth's layers look like in a scale model?

> **TEACHER NOTE** Students may need direction in this summative activity to understand that they will need to create a model to scale. Provide them with some examples of how they can change large distances to small ones and still retain the scale. An example can be a map of distances with the key showing the scale.
>
> This assignment will demonstrate to you if students have understood how the material in Earth's layers can rise and fall depending on its density and, therefore, present an ever-changing surface composition. This can be completed by individual students or in small groups. Using groups gives an opportunity for students to discuss their ideas and defend their reasoning.

- Hands-On Lab: Earth's Interior

TEI Research Report

- Activity: Engineering Design Sheet

STEM Project Starter Growth of the Atmosphere `Recommended 90 minutes`

How have scientists reconstructed the dramatic changes in Earth's atmosphere early in the planet's history, and how did those changes promote terrestrial life?

> **TEACHER NOTE** This summative STEM project allows you to assess how students understand the development of the atmosphere and the importance photosynthesis has played in the development of living things.
>
> The Everts exercise may be completed by individual students. However, the research portion is a nice activity to have students complete in small groups, dividing the research, then coming back to the group to share their knowledge. Divide students into groups so that all events are covered. Have the groups give their presentations to the whole class so that all students gain an understanding of each of the events.

`TEI` **Events**
`TEI` **Research**

Evidence for Plate Tectonics

The Five E Instructional Model

Science Techbook follows the 5E instructional model. This Model Lesson includes strategies for each of the 5Es. As you design the inquiry-based learning experience for students, be sure to collect data during instruction to drive your instructional decisions. Point-of-use teacher notes are also provided within each E-tab.

Engage 45–90 minutes

Engage Media Resources

The resources found in Engage are intended to stimulate students by exposing them to a phenomenon relevant to the content of the lesson. Engage also provides examples of relevant real-world applications that allow students to begin to make observations and relate the science content to their everyday lives. The Core Interactive Text (CIT) and media resources are carefully designed to prompt students to begin asking questions that they can investigate during the Explore phase of the lesson. They should also start collecting evidence to address the Explain question located at the bottom of the Engage page.

> **TEACHER NOTE** **Investigative Phenomenon:** There are numerous examples of fossil evidence and animal group distribution which support the idea that the modern-day separate continents were once joined together. The text and image focus on two examples, those of Mesosaurus (a genus of now extinct aquatic reptiles from the early Permian) and the now-extinct genus of seed fern Glossopteris. These were two groups of organisms identified by Alfred Wegener as evidence for continental drift. Have the students use the text, image, and "A Perplexing Puzzle" video to learn about why these fossils were so important in the development of the theory of continental drift. Can they suggest why this theory was not immediately widely accepted (a lack of an obvious mechanism) despite the evidence provided by Wegener?

- Core Interactive Text: Evaluating Evidence for Plate Tectonics
- Image: Plants Can't Swim
- Video: A Perplexing Puzzle

Explain Question

The Explain question focuses students on gathering information in the Explore section. The Explain question can be used to

- Record what students already know related to the Explain question.
- Serve as a template or model for students to generate their own scientific questions.
- Collect evidence as students work through the lesson.
- Allow students to reflect on their growth before and after the lesson.

Explain

What evidence do scientists have that Earth's crust is divided into moving plates?

- Image: Matching Mesosaurus

Engage Formative Assessment

Technology Enhanced Items (TEIs) found on the Engage page enable you to collect data on students' prior knowledge and identify the common misconceptions they may possess that are related to the topic of study. These items are designed as quick checks for understanding and allow each student one attempt at each question. You can use the data collected to decide whether to assign additional resources to the class, or determine what individual or groups of students may need reinforcement or accelerated learning, prior to completing the Explore portion of the lesson.

TEACHER NOTE Use this student response to evaluate students' prior knowledge of the concept. The Model Lesson provides information on common student misconceptions.

TEI Your Ideas

Before You Begin

What Do I Already Know about the Evidence for Plate Tectonics?

TEACHER NOTE This item is intended to provide the teacher with feedback on students' prior knowledge of this topic. Students should be familiar with the properties of Earth's layers from previous studies. It should be used as a formative assessment.

TEI Earth's Structure

TEACHER NOTE The purpose of this diagnostic formative assessment is to determine student familiarity with the prerequisite concept of convection currents, which are an important part of the theory of how seafloor spreading works. Note that the order of the phrases "Warmed water rises due to lower density" and "Cooler water sinks due to greater density" is interchangeable, as both events happen simultaneously.

TEI Convection

TEACHER NOTE This formative assessment is intended to provide the teacher with feedback on students' prior knowledge of this topic. Its main focus is to determine student familiarity with the various kinds of radioactive decay, some of which are used to provide absolute-age dates of rocks. Such dates are helpful in tracing the history of Earth's plate movements. When discussing this question, be sure to remind students that absolute dates refers to the technique of absolute dating and provide only an approximate age for a rock.

TEI Kinds of Radioactive Decay

- Video: Earth's Layers
- Video: Convection
- Animation: Forms of Radioactivity

Explore `135 minutes`

Lesson Questions (LQs):
1. What is the process of seafloor spreading?
2. What evidence supports the theory of plate tectonics?
3. What are the processes thought to drive the motion of tectonic plates?

Effective science instruction involves a student-centered rather than a teacher-centered approach. This can be accomplished either with Directed Inquiry or Guided Inquiry, depending on the needs and abilities of your class. Encourage students to select a variety of resources in their pursuit of answers as they work through Explore, with the end goal of constructing their scientific explanation in the Explain tab.

Directed Inquiry	Guided Inquiry
In Directed Inquiry, teachers provide students with a sequence of specified resources, challenging questions, and clear outcomes. Within this context students are given the opportunity to interact independently with each resource as prescribed by the teacher. Often different students groups can be guided through several different resources at the same time. For example, one group could work on a reading passage while a second group conducts a small-group Hands-On Activity with the teacher, and a third group is independently engaged with an online interactive resource.	In Guided Inquiry, students have independence to decide the scope and sequence of their investigations. Using resources from Techbook, students determine for themselves which resources they will Explore to answer the Lesson Questions. It is important to note that each student will choose multiple resources, but no one student is expected to use all the resources available. Students also determine the order in which to explore these resources and how to record their findings.

NGSS Components

SEP	CCC
■ Analyzing and Interpreting Data ■ Engaging in Argument from Evidence	■ Patterns ■ Systems and System Models

Lesson Question: What Is the Process of Seafloor Spreading?   `Recommended 45 minutes`

TEACHER NOTE Science and Engineering Practice: Engaging in Argument from Evidence: As students explore this concept, they will be focusing a lot of attention on evidence to support the argument that Earth's crust is composed of moving plates. Suggest to students that they make a T-chart with two headings that they can continually add to over the course of the lesson. Have them label one heading "Hypothesis or Theory" and the other heading "Evidence in Support." Then have them record information on their table as they encounter it in the lesson. At the end of the lesson, have students use their charts to write a claim and compose an explanation of the evidence in support of that claim. Have pairs of students trade papers and critique their partner's claim, evidence, and reasoning.

- Core Interactive Text: What Is the Process of Seafloor Spreading?
- Video: The Lithosphere
- Image: Age of Oceanic Lithosphere
- Video: How Is Magnetism Recorded in Rocks?

TEACHER NOTE Consider having the entire class work on the exploration Evidence for Tectonic Plates as one group. Lead the exploration, asking students for their input along the way. Take advantage of the questions posed in the teacher's guide associated with this exploration, making sure you guide students in a way that will best assist them so they will be prepared to answer these questions on their own.

- Image: Magnetic Striping
- Reading Passage: Exploring Deep-Sea Ecosystems

Formative Assessment:

Throughout Explore, Technology Enhanced Items (TEIs) are embedded as multi-dimensional formative checks for understanding. You can use the data they provide to

- assign additional support
- extend learning
- design additional learning tasks to clarify student misconceptions

The Explore TEIs provide students with three attempts to demonstrate their proficiency. Scaffolded feedback is provided for each attempt. If a student does not achieve proficiency by the third attempt, a media asset is provided as an additional learning opportunity.

TEACHER NOTE **Crosscutting Concepts: Patterns:** This item requires students to analyze empirical evidence in order to observe patterns; specifically, they analyze the pattern of magnetic-field differences along a section of the ocean floor to determine the location of a mid-ocean ridge and the locations of the oldest and youngest rocks. Before assigning this item to students, have them create sketches to model changes in the magnetic-field pattern in the ocean floor as a function of time. This will prepare them for thinking about the data presented in the item and thus be able to interpret it on their own without having to refer to resources. After students have completed this item, gather the class as a whole and ask them to explain their reasoning in completing it. Ask: How do you think your approach to the data was similar to or different from the scientists who first looked at data such as this?

TEI **Magnetism Data**

Lesson Question: What Evidence Supports the
Theory of Plate Tectonics?

Recommended 45 minutes

- Core Interactive Text: What Evidence Supports the Theory of Plate Tectonics?
- Video: Pangaea
- Video: Introducing Continental Drift
- Reading Passage: The Man Behind Continental Drift
- Video: Movement of Tectonic Plates
- Video: Ocean Ridges
- Hands-On Lab: Demonstrating Plate Tectonics

TEACHER NOTE **Science and Engineering Practices: Analyzing and Interpreting Data:** This item helps students compare data related to theories of Earth's continents and examine each piece of evidence for consistency with a theory. As they work through this item, students differentiate between the two fundamental theories that underlie our understanding of how Earth's continents have changed over time. Before assigning this item, discuss with students that it is a common misconception that the continental drift hypothesis is the same as plate-tectonics theory. Ask students to react to this misconception during a class discussion to bring out their ideas. Steer the discussion as needed to assist them in understanding that the two are separate but related ideas. When students have completed the assessment item, gather them together as a class and ask them to discuss why the different pieces of evidence belong to each idea.

TEACHER NOTE **Misconception:** Students may think that the continental drift hypothesis and the theory of plate tectonics are the same idea. In fact, Alfred Wegener proposed the continental drift hypothesis half a century before another generation of geologists outlined the theory of plate tectonics. Continental drift was Wegener's attempt to explain various observations that suggested the continents were once joined together. Plate tectonics is the mechanism by which continental drift happens.

TEI Evidence for Two Ideas

Lesson Question: What Are the Processes Thought to Drive the Motion of Tectonic Plates?

Recommended 45 minutes

- Core Interactive Text: What Are the Processes Thought to Drive the Motion of Tectonic Plates?
- Video: Subduction
- Video: Why Tectonic Plates Move

TEACHER NOTE Crosscutting Concept: Systems and System Models: In completing this item, students consider Earth as a system as they construct a model to describe subduction of oceanic crust under continental crust. As a follow-up activity, ask students to apply plate-tectonic theory to explain why volcanoes tend to form at plate boundaries, why the Hawaiian Islands are a chain with decreasing volcanic activity along the chain, and why the Red Sea is widening. Have students create models on paper and then explain their models by giving an oral presentation or by writing an explanation on paper to hand in. As part of their presentation, have students describe how models of Earth can be useful to describe the behavior of the plates while also noting the limitations of such models due to necessary approximations in models

TEI Plate Tectonics and Subduction

Explore More Resources

Resources in Explore More Resources support differentiation within your classroom by

- providing additional visualization of content
- affording extension of content to those students ready for acceleration
- offering Lexile reading levels for reading passages

Online explorations and hands-on experiences are provided so that students can conduct virtual investigations, collect and design investigations, and collect and analyze data; these skills are essential to developing scientific understanding.

Explain `45–90 minutes`

In Explore, students
1. uncovered scientific understandings
2. conducted investigations
3. analyzed data, text, and other media resources
4. collected evidence to support their scientific explanation

In Explain, provide students with time to formally compose their scientific explanations around the Explain or student-generated questions using evidence collected from Explore.

Scientific explanations are student responses, either written or orally presented, that explain scientific phenomena based upon evidence. Developing a scientific explanation requires students to analyze and interpret data to construct meaning out of the data. There are three main components to the scientific explanation: the claim, the evidence, and the reasoning.

To help students to communicate their scientific explanations, allow them to utilize the multimedia creation tools such as Board Builder and Whiteboard. Remind them that they may upload image, audio, and video files using the "attach file" option to communicate their scientific explanations.

Students may construct their scientific explanations individually or within a small group of students. Students should communicate their explanations with other classmates, and provide constructive criticism and refine their explanations prior to submission to the teacher. If explanations are used as a formative assessment, you can provide additional feedback and comments to support students as they refine their explanations.

EXPLAIN

What evidence do scientists have that Earth's crust is divided into moving plates?

Elaborate with STEM `45–90 minutes`

*Elaborate with STEM are optional extension resources available after students have demonstrated proficiency with standards addressed previously in the concept.

NGSS Components

SEP	CCC
■ Asking Questions and Defining Problems ■ Planning and Carrying Out Investigation ■ Using Mathematics and Computational Thinking ■ Constructing Explanations and Designing Solutions ■ Obtaining, Evaluating, and Communicating Information	■ Cause and Effect ■ Structure and Function

STEM In Action `45 minutes`

STEM in Action ties the scientific concepts to real-world applications, with many connecting to STEM careers. Technology Enhanced Items (TEIs) expect students to critically read the Core Interactive Text (CIT) and review the provided media resources.

Applying Evidence for Plate Tectonics

- Core Interactive Text: Applying Evidence for Plate Tectonics
- Image: Earth's Fiery Boundaries
- Video: Antarctica Rocks!
- Image: Exploring Europa

TEACHER NOTE This summative assessment gives students the opportunity to demonstrate their understanding of how evidence from multiple sources is used to confirm the theory of plate tectonics.

TEI Exploring Europa

STEM Project Starters

STEM Project Starters provide additional real-world contexts that require students to apply and extend their content knowledge related to the concept. STEM Project Starters can also serve as an alternative instructional hook presented at the beginning of the learning progression. The project can then be revisited throughout and at the end of the 5E learning cycle, for students to apply content knowledge.

STEM Project Starter Magnetic Declination

`Recommended 45 minutes`

What is the difference between magnetic north and geographic north?

TEACHER NOTE This STEM project relates to magnetic declination, or the difference between the magnetic and geographic poles, as well as the importance of considering this difference when using an important geologic device, the Brunton compass.

- Exploration: Evidence for Plate Tectonics
- Video Magnetic Declination

TEI Compass Correction

TEI Losing Your Way?

STEM Project Starter Patterns in the Seafloor

`Recommended 45 minutes`

Can you predict normal and reverse polarity of ocean-floor rocks?

TEACHER NOTE This formative assessment requires students to interpret geologic data both quantitatively and qualitatively in order to make predictions. They must apply their understanding of the formation of a mid-ocean ridge and its geology while demonstrating spatial and mathematical reasoning.

- Image: Polarity Change

TEI Patterns in the Seafloor

Evaluate `45–90 minutes`

Explain Question:
What evidence do scientists have that Earth's crust is divided into moving plates?

Lesson Questions (LQ):
- What is the process of seafloor spreading?
- What evidence supports the theory of plate tectonics?
- What are the processes thought to drive the motion of tectonic plates?

Throughout instruction and the 5E learning cycle, you will have collected formative assessment data to drive the assignment of resources and experiences to students. Evaluate is intended to include summative assessment checks for proficiency. You can use the Explain and Lesson Questions for the concept as a summative assessment in a variety of ways such as these:

- Post each Lesson Question (LQ) in various locations in the classroom, and have small groups of students generate claim statements related to the Lesson Question (LQ). Other students can add to the claim, or refute the claim, during a gallery walk where they place additional pieces of evidence on each Lesson Question (LQ) poster.
- Assign small groups of students to each Lesson Question (LQ) and have the groups generate a poster, board, graphic, or piece of text that answers the question. Use a jigsaw approach and create a second set of groups that contain members from each Lesson Question (LQ) group to share their ideas.
- Ask students to return to their initial ideas for the Explain question and add additional details and evidence.

Encourage students to review the concept review and complete the Student Self-Check practice assessment prior to assigning the Summative Teacher Concept assessment.

- Student Review and Practice Assessment
- Teacher Concept Assessments

Relative Dating

The Five E Instructional Model

Science Techbook follows the 5E instructional model. This Model Lesson includes strategies for each of the 5Es. As you design the inquiry-based learning experience for students, be sure to collect data during instruction to drive your instructional decisions. Point-of-use teacher notes are also provided within each E-tab.

Engage 45–90 minutes

Engage Media Resources

The resources found in Engage are intended to stimulate students by exposing them to a phenomenon relevant to the content of the lesson. Engage also provides examples of relevant real-world applications that allow students to begin to make observations and relate the science content to their everyday lives. The Core Interactive Text (CIT) and media resources are carefully designed to prompt students to begin asking questions that they can investigate during the Explore phase of the lesson. They should also start collecting evidence to address the Explain question located at the bottom of the Engage page.

> **TEACHER NOTE** **Investigative Phenomenon:** To get students thinking about how they can determine the relative age of something, post images of several people around the room. You could include photographs of other teachers, celebrities, or random people obtained by searching in the Discovery Education library. Be sure they span different generations. Have students walk around and try to put them in order by age. Ask students if they needed the exact age of each person in order to get their order correct. How might scientists be able to determine the relative age of rocks, fossils, and events without knowing the exact age?

- Core Interactive Text: Thinking about Relative Age
- Image: Family Gathering
- Video: Rock Formations and Fault Lines
- Video: America's Tectonic Foundation
- Video: The K-Pg Layer

Explain Question

The Explain question focuses students on gathering information in the Explore section. The Explain question can be used to

- Record what students already know related to the Explain question.
- Serve as a template or model for students to generate their own scientific questions.
- Collect evidence as students work through the lesson.
- Allow students to reflect on their growth before and after the lesson.

Explain
What is relative dating, and how can it be determined?

- Image: Promicrocerasplanicosta, Early Jurassic

Engage Formative Assessment
Technology Enhanced Items (TEIs) found on the Engage page enable you to collect data on students' prior knowledge and identify the common misconceptions they may possess that are related to the topic of study. These items are designed as quick checks for understanding and allow each student one attempt at each question. You can use the data collected to decide whether to assign additional resources to the class, or determine what individual or groups of students may need reinforcement or accelerated learning, prior to completing the Explore portion of the lesson.

> **TEACHER NOTE** Use responses to this formative assessment to evaluate students' prior knowledge of the relative age dating. Have students work in think-pair-share groups to discuss their thoughts about fossil location, fossil age, and rock layer formation.
>
> Some students may mention that absolute age dating is also used; it should be reinforced that in most situations, relative age dating is used because absolute dating is not possible due to the type of rock.

TEI Fossils in Mountains

Before You Begin
What Do I Already Know about Relative Dating?

> **TEACHER NOTE** Use this formative assessment to evaluate students' prior knowledge of determining relative age. Suggested use includes students working in think-pair-share groups to discuss their thoughts about why the oldest rock layers are at the bottom and what impact the fault had on the results. Teachers should note that the answers should be in the order given.

TEI Determining Relative Age

> **TEACHER NOTE** This formative assessment is intended to provide the teacher with feedback on students' prior knowledge about the relative age dating. It is suggested that after individuals complete the assessment, have a class discussion activity with your students and then have the students write a response.

TEI Relative Age Dating

TEACHER NOTE This formative assessment is intended to provide the teacher with feedback on student misconceptions and prior knowledge of the principles and laws of relative age dating. It is designed to be used at the beginning of the lesson so that the teacher has an understanding of the content to review or cover with the students.

It is suggested that after individuals complete the assessment, it can be completed as a think-pair-share activity with your students and then have the students write a response.

TEI **Principles and Laws**

- Video: Why Is the Study of Rocks Important?
- Video: Three Types of Rocks
- Video: The Grand Canyon

Explore `135 minutes`

Lesson Questions (LQs):

1. What broad, underlying assumptions or philosophies do scientists use to explain Earth's history?
2. What dating techniques do scientists use to determine the relative ages of rocks?
3. How do scientists describe gaps in the rock record?

Effective science instruction involves a student-centered rather than a teacher-centered approach. This can be accomplished either with Directed Inquiry or Guided Inquiry, depending on the needs and abilities of your class. Encourage students to select a variety of resources in their pursuit of answers as they work through Explore, with the end goal of constructing their scientific explanation in the Explain tab.

Directed Inquiry	Guided Inquiry
In Directed Inquiry, teachers provide students with a sequence of specified resources, challenging questions, and clear outcomes. Within this context students are given the opportunity to interact independently with each resource as prescribed by the teacher. Often different students groups can be guided through several different resources at the same time. For example, one group could work on a reading passage while a second group conducts a small-group Hands-On Activity with the teacher, and a third group is independently engaged with an online interactive resource.	In Guided Inquiry, students have independence to decide the scope and sequence of their investigations. Using resources from Techbook, students determine for themselves which resources they will Explore to answer the Lesson Questions. It is important to note that each student will choose multiple resources, but no one student is expected to use all the resources available. Students also determine the order in which to explore these resources and how to record their findings.

NGSS Components

SEP	CCC
■ Developing and Using Models ■ Analyzing and Interpreting Data ■ Constructing Explanations and Designing Solutions ■ Engaging in Argument from Evidence ■ Obtaining, Evaluating, and Communicating Information	■ Patterns ■ Cause and Effect

Lesson Question: What Broad, Underlying Assumptions or Philosophies Do Scientists Use to Explain Earth's History?

`Recommended 45 minutes`

TEACHER NOTE **Practices: Science and Engineering Practice: Engaging in Argument from Evidence:** In this concept, students will read scientific text critically in order to learn how to relatively date geological formations and why it is so important. After obtaining this information, students will summarize information about how the age of rocks provides important clues about how Earth's history has been affected. Students may think that geologists usually go inside caves or dig deep tunnels to carry out their studies and investigate the Earth's history, whereas all geological formations, such as folds and faults, are important information sources. Take the students to observe some real geologic formations in the neighborhood and ask them to formulate hypotheses about the geological history of the region.

- Core Interactive Text: What Broad, Underlying Assumptions or Philosophies Do Scientists Use to Explain Earth's History?
- Image Ranger Talking About Dinosaur Bones
- Video: 65 Million Years Ago: Asteroid and Climate Change
- Video: Uniformitarianism and Other Ideas

TEACHER NOTE **Practices: Science and Engineering Practice: Analyzing and Interpreting Data:** This Hands-On Activity develops the students' ability to analyze data in order to make valid claims distinguishing between the philosophies of uniformitarianism and catastrophism. To help the students visualize the effects of catastrophic events on geological formations, organize a demonstration inspired by the video Stratigraphic Principles. Stress a simulated sedimentary formation in various ways. The students must express their observations and describe the results in terms of uniformitarianism and catastrophism.

- Hands-On Activity: Distinguishing between Catastrophe and Uniformity

Formative Assessment:

Throughout Explore, Technology Enhanced Items (TEIs) are embedded as multi-dimensional formative checks for understanding. You can use the data they provide to

- assign additional support
- extend learning
- design additional learning tasks to clarify student misconceptions

The Explore TEIs provide students with three attempts to demonstrate their proficiency. Scaffolded feedback is provided for each attempt. If a student does not achieve proficiency by the third attempt, a media asset is provided as an additional learning opportunity.

TEACHER NOTE Practices: Science and Engineering Practice: Constructing Explanations and Designing Solutions: This formative assessment requires the students to construct explanations about Earth's history by applying scientific ideas, principles, and evidence. In particular, the students must associate different processes and events with the philosophies of catastrophism and uniformitarianism. To improve the argumentation skills of the students, engage them in a role play and divide the class into three groups. Two groups have to support the two opposite philosophies, and the third one must explain why a more inclusive explanation would be better.

TEI Earth's History

Lesson Question: What Dating Techniques Do Scientists Use to Determine the Relative Ages of Rocks?

Recommended 55 minutes

TEACHER NOTE Misconception: Students may believe that scientists use only absolute dating techniques to determine the age of ancient items and events—for example, students may believe that all rocks and fossils that originated in Earth's past have been radiometrically dated. In fact, scientists also use relative dating techniques, which are used to determine the relative order of past events.

- Core Interactive Text: What Dating Techniques Do Scientists Use to Determine the Relative Ages of Rocks?

TEACHER NOTE Connections: Crosscutting Concept: Cause and Effect: In this Exploration, students experience the relative dating approach used by geologists. They can observe empirical data in order to make claims about the cause-and-effect relationships between different types of stresses and the structure of a geological formation. Absolute dating techniques are employed in geological studies, too; therefore, it is important that students understand the difference between the types of answers achievable with these two approaches. Combine this activity with the Exploration Absolute Dating, in Explore More Resources, to help the students clarify the differences. Then, let the students discuss in groups how these two types of dating can be combined to obtain more reliable information.

TEACHER NOTE Misconception: Students may believe that the layers of material that cover ancient rocks and fossils—especially in high-elevation areas or in areas where the terrain is flat and no mountains or hills are nearby—originated from outer space. In fact, although it is true that Earth accumulates debris from meteorites, most of the material that covers ancient rocks and fossils resulted from erosion and deposition processes on Earth. Erosion and transportation processes can carry sediment great distances before it is deposited—e.g., material from the Rocky Mountains eventually can be deposited on flood plains within the Mississippi River Valley. Also, low-lying regions of deposition eventually can be uplifted to higher elevations.

- Exploration: Relative Dating
- Video: Stratigraphic Principles
- Video: Folding and Tilting
- Image: Sedimentary Rocks
- Image: Igneous Rocks
- Hands-On Lab: Determining the Relative Ages of Events

TEACHER NOTE Practices: Science and Engineering Practice: Developing and Using Models: This formative assessment develops students' ability to use evidence-based geological models as basic tools to interpret past events and make predictions and the natural processes affecting the shape of the Earth's surface. To extend this activity, divide the students into small groups and assign them one of the pictures to analyze. The students must apply the techniques recognized to relatively date the various components of the geological formation.

TEI Relative Dating

- Image: Conglomerate
- Image: Lunar Breccia

Lesson Question: How Do Scientists Describe Gaps in the Rock Record?

Recommended 35 minutes

- Core Interactive Text: How Do Scientists Describe Gaps in the Rock Record?
- Image: Unconformity
- Video: Unconformities

TEACHER NOTE Connections: Crosscutting Concept: Patterns: As a general strategy, you can integrate the formative activities and materials by concluding the lesson with an excursion to a nearby geologic formation. Find a geologically relevant area that presents at least one type of unconformity and that allows applying a minimum of two relative dating techniques. Divide the students into small groups. The students must collect data, take pictures, and write down their preliminary observation. Then, ask them to use the patterns they observe in these geologic systems to attempt a relative dating of the geological formation. As follow-up homework, assign students to write a report about their experience.

- Reading Passage: Grand Canyon Adventure
- Image: Three Unconformities
- Image: Angular Unconformity
- Image: Disconformity
- Image: Nonconformity

UNIT 3: Evolving Earth

TEACHER NOTE Practices: Science and Engineering Practice: Obtaining, Evaluating, and Communicating Information: This formative assessment makes the students critically read a scientific passage covering the concepts treated in this lesson. In particular, the students must evaluate multiple scientific claims about applying the relative dating principles and understand how the geologists describe gaps in the rock record. To recognize the word incorrectly used in the text, the students must develop their reasoning skills and identify the inconsistent passages. Extend the activity by asking the students to propose the correct words to use in the text and to justify their choices.

TEI Geological Studies

Explore More Resources

Resources in Explore More Resources support differentiation within your classroom by

- providing additional visualization of content
- affording extension of content to those students ready for acceleration
- offering Lexile reading levels for reading passages

Online explorations and hands-on experiences are provided so that students can conduct virtual investigations, collect and design investigations, and collect and analyze data; these skills are essential to developing scientific understanding.

Explain `45–90 minutes`

In Explore, students
1. uncovered scientific understandings
2. conducted investigations
3. analyzed data, text, and other media resources
4. collected evidence to support their scientific explanation

In Explain, provide students with time to formally compose their scientific explanations around the Explain or student-generated questions using evidence collected from Explore.

Scientific explanations are student responses, either written or orally presented, that explain scientific phenomena based upon evidence. Developing a scientific explanation requires students to analyze and interpret data to construct meaning out of the data. There are three main components to the scientific explanation: the claim, the evidence, and the reasoning.

To help students to communicate their scientific explanations, allow them to utilize the multimedia creation tools such as Board Builder and Whiteboard. Remind them that they may upload image, audio, and video files using the "attach file" option to communicate their scientific explanations.

Students may construct their scientific explanations individually or within a small group of students. Students should communicate their explanations with other classmates, and provide constructive criticism and refine their explanations prior to submission to the teacher. If explanations are used as a formative assessment, you can provide additional feedback and comments to support students as they refine their explanations.

EXPLAIN
What is relative dating, and how can it be determined?

Elaborate with STEM `45–135 minutes`

*Elaborate with STEM are optional extension resources available after students have demonstrated proficiency with standards addressed previously in the concept.

NGSS Components

SEP	CCC
■ Asking Questions and Defining Problems ■ Developing and Using Models ■ Analyzing and Interpreting Data ■ Constructing Explanations and Designing Solutions ■ Engaging in Argument from Evidence ■ Obtaining, Evaluating, and Interpreting Information	■ Patterns ■ Cause and Effect ■ Systems and System Models ■ Structure and Function ■ Stability and Change

STEM In Action `45 minutes`

STEM in Action ties the scientific concepts to real-world applications, with many connecting to STEM careers. Technology Enhanced Items (TEIs) expect students to critically read the Core Interactive Text (CIT) and review the provided media resources.

Applying Relative Dating

- Core Interactive Text: Applying Relative Dating
- Video: The Fossil Record
- Video: Science in Progress: How Complete Is the Fossil Record?
- Image: Oil Workers Looking at a Geologic Cross Section
- Image: Ordovician
- Image: Microfossils

TEACHER NOTE This activity may be used as a summative assessment to determine how well students understand the role of rock formation appearance on determining Earth's history. Students will assess the best way to collect data about the two rock formations so that similarities and differences about the structures are identified. It is suggested that students complete the assessment independently.

TEI Comparing Formations

STEM Project Starters

STEM Project Starters provide additional real-world contexts that require students to apply and extend their content knowledge related to the concept. STEM Project Starters can also serve as an alternative instructional hook presented at the beginning of the learning progression. The project can then be revisited throughout and at the end of the 5E learning cycle, for students to apply content knowledge.

STEM Project Starter: Using Relative Dating to Locate Oil
Recommended 135 minutes

How does knowing the age of the rock assist in locating the presence of petroleum?

> **TEACHER NOTE** This summative assessment gives students the opportunity to apply their knowledge of relative age dating to a real-life situation. It is suggested that this assessment be used as a group activity, and the project plan be shared with another group for peer review.

TEI Research Report

STEM Project Starter Rock Profile Model
Recommended 90 minutes

How do index fossils help to determine the ages of rock layers in a profile?

> **TEACHER NOTE** This summative assessment gives students the opportunity to demonstrate their knowledge of relative age dating by constructing an accurate model of a sedimentary rock profile, including index fossils, various rock types, and environmental conditions to explain the history of their rock model. Rock profile models may be physically constructed from craft materials, or they may be posters or interactive digital models, depending on available materials and equipment and student inclination. Student groups may compare completed models to attempt to line up the corresponding layers of multiple models, based on the presence of index fossils.

TEI Descriptions
TEI Calculations
TEI Image or Diagram

Evaluate `45–90 minutes`

Explain Question:
What is relative dating, and how can it be determined?

Lesson Questions (LQ):
- What broad, underlying assumptions or philosophies do scientists use to explain Earth's history
- What dating techniques do scientists use to determine the relative ages of rocks?
- How do scientists describe gaps in the rock record?

Throughout instruction and the 5E learning cycle, you will have collected formative assessment data to drive the assignment of resources and experiences to students. Evaluate is intended to include summative assessment checks for proficiency. You can use the Explain and Lesson Questions for the concept as a summative assessment in a variety of ways such as these:

- Post each Lesson Question (LQ) in various locations in the classroom, and have small groups of students generate claim statements related to the Lesson Question (LQ). Other students can add to the claim, or refute the claim, during a gallery walk where they place additional pieces of evidence on each Lesson Question (LQ) poster.
- Assign small groups of students to each Lesson Question (LQ) and have the groups generate a poster, board, graphic, or piece of text that answers the question. Use a jigsaw approach and create a second set of groups that contain members from each Lesson Question (LQ) group to share their ideas.
- Ask students to return to their initial ideas for the Explain question and add additional details and evidence.

Encourage students to review the concept review and complete the Student Self-Check practice assessment prior to assigning the Summative Teacher Concept assessment.

- Student Review and Practice Assessment
- Teacher Concept Assessments

Evidence for Evolution

The Five E Instructional Model
Science Techbook follows the 5E instructional model. This Model Lesson includes strategies for each of the 5Es. As you design the inquiry-based learning experience for students, be sure to collect data during instruction to drive your instructional decisions. Point-of-use teacher notes are also provided within each E-tab.

Engage   45–90 minutes

Engage Media Resources
The resources found in Engage are intended to stimulate students by exposing them to a phenomenon relevant to the content of the lesson. Engage also provides examples of relevant real-world applications that allow students to begin to make observations and relate the science content to their everyday lives. The Core Interactive Text (CIT) and media resources are carefully designed to prompt students to begin asking questions that they can investigate during the Explore phase of the lesson. They should also start collecting evidence to address the Explain question located at the bottom of the Engage page.

> **TEACHER NOTE** **Investigative Phenomenon:** Prior to working through Engage, have students brainstorm how similar they are to a number of familiar organisms. Collect their ideas in a table with a column for each organism, and have them suggest features that are shared with humans. Select organisms with a wide range of similarities and some that are very different from humans. Provide images to help them make their comparisons (example images can be found in Explore More Resources and include the gorilla, horse, Canada goose, collared lizard, and spruce tree.) Upon completion, ask students whether they think, sometime in the distant past, they could have had an ancestor in common with each of the organisms. If so, which ones do they think they are most closely related to, and which are probably the most distant relations?

- Core Interactive Text: Everything is a Relative
- Image: A Relatively Close Relative
- Video: The Rise of the Primates
- Image: A Primate Cladogram or Tree of Life
- Image: Hominid Skulls from East Africa
- Video: The Evolution of Genus Homo
- Video: The Discovery of Lucy
- Video: The Origins of Bipedalism

Explain Question

The Explain question focuses students on gathering information in the Explore section. The Explain question can be used to

- Record what students already know related to the Explain question.
- Serve as a template or model for students to generate their own scientific questions.
- Collect evidence as students work through the lesson.
- Allow students to reflect on their growth before and after the lesson.

Explain

How do scientists use evidence from a wide variety of sources to determine evolutionary relationships?

- Image: Galapagos Marine Iguana

Engage Formative Assessment

Technology Enhanced Items (TEIs) found on the Engage page enable you to collect data on students' prior knowledge and identify the common misconceptions they may possess that are related to the topic of study. These items are designed as quick checks for understanding and allow each student one attempt at each question. You can use the data collected to decide whether to assign additional resources to the class, or determine what individual or groups of students may need reinforcement or accelerated learning, prior to completing the Explore portion of the lesson.

TEACHER NOTE Use this student response to evaluate students' prior knowledge of the basic principle of evolution, and their ability to cite evidence in support of their claim. The Model Lesson provides information on common student misconceptions.

TEI Iguana Evolution

Before You Begin

What Do I already know about evidence for evolution?

TEACHER NOTE This formative assessment item is intended to provide the teacher with feedback on how well students can interpret evidence from fossils.

TEI Largest Increase?

TEI Analyzing the Data

- Video: Alligator Evolution
- Video: Heredity
- Image: Mutations and Changes in Traits

Explore   135 minutes

Lesson Questions (LQs):

1. How are fossils used as evidence for evolution?
2. How do scientists explain gaps and sudden appearances in the fossil record?
3. What do morphological and developmental homologies tell us about common ancestry?
4. How do molecular sequences provide evidence for evolution?
5. How does biogeography provide evidence for evolution?
6. Can evolution be observed today?

Effective science instruction involves a student-centered rather than a teacher-centered approach. This can be accomplished either with Directed Inquiry or Guided Inquiry, depending on the needs and abilities of your class. Encourage students to select a variety of resources in their pursuit of answers as they work through Explore, with the end goal of constructing their scientific explanation in the Explain tab.

Directed Inquiry	Guided Inquiry
In Directed Inquiry, teachers provide students with a sequence of specified resources, challenging questions, and clear outcomes. Within this context students are given the opportunity to interact independently with each resource as prescribed by the teacher. Often different students groups can be guided through several different resources at the same time. For example, one group could work on a reading passage while a second group conducts a small- group Hands-On Activity with the teacher, and a third group is independently engaged with an online interactive resource.	In Guided Inquiry, students have independence to decide the scope and sequence of their investigations. Using resources from Tech book, students determine for themselves which resources they will Explore to answer the Lesson Questions. It is important to note that each student will choose multiple resources, but no one student is expected to use all the resources available. Students also determine the order in which to explore these resources and how to record their findings.

NGSS Components

SEP	CCC
■ Constructing Explanations and Designing Solutions ■ Developing and Using Models Engaging in Argument from Evidence ■ Obtaining, Evaluating and Communicating Information	■ Cause and Effect ■ Patterns ■ Structure and Function

Lesson Question: How are Fossils Used as Evidence for Evolution?

`Recommended 90 minutes`

TEACHER NOTE If students have already encountered a more detailed description of the fossil record in the concept "The History of Life on Earth," then have students use this lesson question for a quick review prior to proceeding with the remainder of the concept.

As students read and comprehend complex texts, view the videos, and complete the interactives, labs, and other Hands-on Activities, have them summarize and obtain scientific and technical information. Students will use this evidence to support their initial ideas on how to answer the Explain Question or their own question they generated during Engage. Have students record their evidence using "My Notebook."

TEACHER NOTE Misconception: Many students think that evolution is something that occurs slowly and only occurred in the past. Although the evolution of new species is often on a timescale of thousands or millions of years, the process is ongoing and can also be very rapid.

- Core Interactive Text: How are fossils used as evidence for evolution?
- Image: Fossils
- Video: The Formation of Fossils
- Video: Evolution and Fossils
- Activity: Hands-On Lab: Describing, Interpreting, and Identifying Fossils
- Video: The Fossil Record
- Image: Trace Fossils
- Video: The Origin of Whales

Lesson Question: How Do Scientists Explain Gaps and Sudden Appearances in the Fossil Record?

`Recommended 45 minutes`

TEACHER NOTE Misconception: Some students may think that because the fossil record is incomplete, it does not provide sufficient evidence that life evolved on Earth over millions of years. Use the analogy of watching a movie to explain how even when you leave the room to go for a snack, upon your return you can usually still follow the plot and guess what happened.

- Core Interactive Text: How do scientists explain gaps and sudden appearances in the fossil record?
- Video: How Complete Is the Fossil Record?
- Video: The Evolutionary Big Bang
- Reading Passage: The Cambrian Explosion

TEACHER NOTE **Misconception:** Some students may erroneously consider evolution as not supported by fossil evidence because there are so many "missing species," so-called "missing links." As stated earlier, the fossil record is not complete. However, numerous intermediate species have been discovered often in location predicted by paleontologists. In addition, the hypothesis of punctuated equilibria provides an explanation for rapid change and the appearance of new species with an absence of intermediate fossils.

TEACHER NOTE After reading the text and viewing the video "Punctuated Equilibrium and Speciation," have students draw a flow chart or a series of simple pictures that model how punctuated equilibria explain the rapid appearance of new forms of organisms in the fossil record.

- Video: Punctuated Equilibrium and Speciation

Lesson Question: What do Morphological and Developmental Homologies Tell Us About Common Ancestry?

Recommended 45 minutes

TEACHER NOTE **Science and Engineering Practices: Engaging in Argument from Evidence:** Before having students read this section, consider having them view the diagram "Forelimb Homologies on Mammals," and write a list of the similarities between them and one unique specialization for each. Use this to stimulate discussion on why this supports the idea that, although they are all different, they probably derived from a common ancestor.

- Image: Forelimb Homologies on Mammals
- Video: What is Homology?
- Video: What is Embryonic Fetal Development?
- Video: Evolution and Embryos of Chordates?
- Video: What is a Developmental Homology?

Formative Assessment:

Throughout Explore, Technology Enhanced Items (TEIs) are embedded as multi-dimensional formative checks for understanding. You can use the data they provide to

- assign additional support
- extend learning
- design additional learning tasks to clarify student misconceptions

The Explore TEIs provide students with three attempts to demonstrate their proficiency. Scaffolded feedback is provided for each attempt. If a student does not achieve proficiency by the third attempt, a media asset is provided as an additional learning opportunity.

TEACHER NOTE **Science and Engineering Practices: Engaging in Argument from Evidence:** In this item, students evaluate evidence of the evolutionary relationship between two organisms. They construct, use, and present a written argument or counterarguments based on data and evidence. To support literacy, help them consider both why the evidence may support a common ancestry and why it may not support a common ancestry by having them draw a pro and con chart. Have them consider several reasons why the evidence is supportive and why the evidence is not supportive of their conclusion.

TEI Bats and Birds

Lesson Question: How Do Molecular Sequences Provide Evidence for Evolution?

Recommended 45 minutes

TEACHER NOTE **Connections: Crosscutting Concepts: Structure and Function Data:** Students may not immediately make the connection between developmental homologies, morphological homologies, and genes and the proteins they code for. Upon completion of this lesson question, emphasize that structural homologies that manifest themselves in organisms are a direct result of mutation in the genes that change the structure of the enzymes that produce the structural changes.

- Core Interactive Text: How Do Molecular Sequences Provide Evidence for Evolution?
- Video: Molecular Comparisons
- Video: Molecular Clocks

TEACHER NOTE **Science and Engineering Practices: Engaging in Argument from Evidence:** In this item, students must interpret a graph about molecular evidence for evolution. Upon completion, initiate discussion about how their conclusions match (or do not match) their original ideas about some of these organisms. For example, most students think rabbits are rodents, so why are they so far apart in terms of diverging from one another?

TEI Molecular Clocks

Lesson Question: How Does Biogeography Provide Evidence for Evolution?

Recommended 45 minutes

TEACHER NOTE **Connections: Crosscutting Concepts: Structure and Function:** Although most examples of adaptive radiation come from small islands, such as those of the Hawaiian and Galapagos archipelagoes, large scale changes in the geography of continents as a result of plate tectonics have produced some spectacular examples of adaptive radiation on continents and other large landmasses. Have students conduct research to find some of these macro examples (Madagascar and Australia are two), and identify organisms that have filled niches occupied by other totally unrelated organisms in other parts of the world. Have students speculate what happens when these once-isolated groups of organisms are brought into contact with competitors from larger continental land masses. What impacts might this have on biodiversity?

- Core Interactive Text: How does biogeography provide evidence for evolution?
- Video: Adaptive Radiation and Biogeography

Lesson Question: Can Evolution Be Observed Today? **Recommended 45 minutes**

TEACHER NOTE Misconception: Some students may think that evolution has occurred but is not occurring anymore. This misconception is partly derived from the emphasis on the fossil record as a source of evidence for evolution having occurred. In fact, evolution is a continual process that does not stop. Changing environmental conditions are continually altering selection pressures on populations, and gene pools are continually changing due to mutation, drift, and gene flow.

- Core Interactive Text: Can Evolution Be Observed Today?
- Image: Bill Depth in Galapagos Medium Ground Finches before Drought
- Image: Bill Depth in Galapagos Medium Ground Finches after Drought?
- Image: Finch Bill Size and Survival
- Video: Bacteria and Antibiotics-Evolution in Action

TEACHER NOTE Science and Engineering Practices: Obtaining, Evaluating, and Communicating Information: In this item, students interpret data on the evolution of herbicide-resistant weeds. Have students share their responses, and then ask them whether they can find examples of resistance to pesticides. Have them share their findings. Ask the question: Why does the application of more pesticides (at higher concentrations or over wider areas) usually exacerbate the problem of resistance?

TEI Resistance to Herbicides

Explore More Resources

Resources in Explore More Resources support differentiation within your classroom by

- providing additional visualization of content
- affording extension of content to those students ready for acceleration
- offering Lexile reading levels for reading passages

Online explorations and hands-on experiences are provided so that students can conduct virtual investigations, collect and design investigations, and collect and analyze data; these skills are essential to developing scientific understanding.

Explain `45–90 minutes`

In Explore, students
1. uncovered scientific understandings
2. conducted investigations
3. analyzed data, text, and other media resources
4. collected evidence to support their scientific explanation

In Explain, provide students with time to formally compose their scientific explanations around the Explain or student-generated questions using evidence collected from Explore.

Scientific explanations are student responses, either written or orally presented, that explain scientific phenomena based upon evidence. Developing a scientific explanation requires students to analyze and interpret data to construct meaning out of the data. There are three main components to the scientific explanation: the claim, the evidence, and the reasoning.

To help students to communicate their scientific explanations, allow them to utilize the multimedia creation tools such as Studio and Whiteboard. Remind them that they may upload image, audio, and video files using the "attach file" option to communicate their scientific explanations.

Students may construct their scientific explanations individually or within a small group of students. Students should communicate their explanations with other classmates, and provide constructive criticism and refine their explanations prior to submission to the teacher. If explanations are used as a formative assessment, you can provide additional feedback and comments to support students as they refine their explanations.

EXPLAIN

How do scientists use evidence from a wide variety of sources to determine evolutionary relationships?

Elaborate with STEM `45–135 minutes`

*Elaborate with STEM are optional extension resources available after students have demonstrated proficiency with standards addressed previously in the concept.

NGSS Components

SEP	CCC
■ Asking Questions and Defining Problems ■ Analyzing and Interpreting Data	■ Developing and Using Models ■ Patterns ■ Structure and Function

STEM In Action `45 minutes`

STEM in Action ties the scientific concepts to real-world applications, with many connecting to STEM careers. Technology Enhanced Items (TEIs) expect students to critically read the Core Interactive Text (CIT) and review the provided media resources.

Applying Evidence for Evolution

TEACHER NOTE You may wish to lead a class discussion about the competing ideas regarding how life originated on Earth. Have students list the evidence for and against each idea.

- Core Interactive Text: Applying Evidence for Evolution
- Video: Underwater Volcanic Nurseries
- Image: Black Smoker

TEACHER NOTE You may wish to have students check PubMed themselves to see that anyone can access information about the human genome. Discuss how this knowledge can be used to trace the history of life (for example, through comparing the relatedness of close and distant species).

- Video: The Human Genome Project

TEACHER NOTE Students complete the Hands-On Laboratory: Analyzing Biological Diversity, Extinction, and Evolution.

- Hands-On Lab: Analyzing Biological Diversity, Extinction, and Evolution

STEM Project Starters

STEM Project Starters provide additional real-world contexts that require students to apply and extend their content knowledge related to the concept. STEM Project Starters can also serve as an alternative instructional hook presented at the beginning of the learning progression. The project can then be revisited throughout and at the end of the 5E learning cycle, for students to apply content knowledge.

STEM Project Starter Fossil Heads   `Recommended 45 minutes`

How can you use two-way tables to analyze morphological traits?

- Core Interactive text: How can you use two-way tables to analyze morphological traits?
- Image: Idealized Morphology of Trilobites and Helemtiids
- Image: Non-Oblate Head Shield
- Video: DE Academy Two-Way Table

TEI Saber-Toothed Cats

STEM Project Starter Evidence for Evolution – Half-Life   `Recommended 45 minutes`

What techniques do scientists use to determine the age of fossils?

- Image: Carbon Dating

TEACHER NOTE This STEM activity reinforces previous learning about radioactive decay, half-life, and fossil dating. It should be used to assess students' knowledge of these topics. Have students work together to answer the questions and then go over the correct solutions as a class.

TEI Evidence for Evolution

STEM Project Starter Plant Classification   `Recommended 45 minutes`

What new structures appeared as plants evolved, and how did those structures help new plant species to succeed?

- Image: Plant Cladogram

TEACHER NOTE This summative assessment is intended to determine how well students can interpret a cladogram and conduct research using the Internet.

Students work in small groups to answer the questions, then share in group discussion.

TEI Structures and Functions

Evaluate 45–90 minutes

Explain Question:
How do scientists use evidence from a wide variety of sources to determine evolutionary relationships?

Lesson Questions (LQ):
- How are fossils used as evidence for evolution?
- How do scientists explain gaps and sudden appearances in the fossil record?
- What do morphological and developmental homologies tell us about common ancestry?
- How do molecular sequences provide evidence for evolution?
- How does biogeography provide evidence for evolution?
- Can evolution be observed today?

Throughout instruction and the 5E learning cycle, you will have collected formative assessment data to drive the assignment of resources and experiences to students. Evaluate is intended to include summative assessment checks for proficiency. You can use the Explain and Lesson Questions for the concept as a summative assessment in a variety of ways such as these:

- Post each Lesson Question (LQ) in various locations in the classroom, and have small groups of students generate claim statements related to the Lesson Question (LQ). Other students can add to the claim, or refute the claim, during a gallery walk where they place additional pieces of evidence on each Lesson Question (LQ) poster.
- Assign small groups of students to each Lesson Question (LQ) and have the groups generate a poster, board, graphic, or piece of text that answers the question. Use a jigsaw approach and create a second set of groups that contain members from each Lesson Question (LQ) group to share their ideas.
- Ask students to return to their initial ideas for the Explain question and add additional details and evidence.

Encourage students to review the concept review and complete the Student Self-Check practice assessment prior to assigning the Summative Teacher Concept assessment.

- Student Review and Practice Assessment
- Teacher Concept Assessments

Mechanisms for Evolution

The Five E Instructional Model
Science Techbook follows the 5E instructional model. This Model Lesson includes strategies for each of the 5Es. As you design the inquiry-based learning experience for students, be sure to collect data during instruction to drive your instructional decisions. Point-of-use teacher notes are also provided within each E-tab.

Engage   45–90 minutes

Engage Media Resources
The resources found in Engage are intended to stimulate students by exposing them to a phenomenon relevant to the content of the lesson. Engage also provides examples of relevant real-world applications that allow students to begin to make observations and relate the science content to their everyday lives. The Core Interactive Text (CIT) and media resources are carefully designed to prompt students to begin asking questions that they can investigate during the Explore phase of the lesson. They should also start collecting evidence to address the Explain question located at the bottom of the Engage page.

> **TEACHER NOTE** **Investigative Phenomenon: Ask the questions:** Why do organisms evolve? How do they evolve? Although many individuals are aware that evolution has (and is) taking place, they may have little understanding of how inevitable the process is. A common misunderstanding is that it is random process that occurs of on geological timescales. Use the video "How Do Species Change over Time?" to initiate thinking about natural selection.

- Core Interactive Text: Thinking About Evolution
- Video: How Do Species Change Over Time?

Explain Question
The Explain question focuses students on gathering information in the Explore section. The Explain question can be used to

- Record what students already know related to the Explain question.
- Serve as a template or model for students to generate their own scientific questions.
- Collect evidence as students work through the lesson.
- Allow students to reflect on their growth before and after the lesson.

Explain
How does evolution work to create the diversity of life on Earth?

- Image: Giraffe Evolution

Engage Formative Assessment

Technology Enhanced Items (TEIs) found on the Engage page enable you to collect data on students' prior knowledge and identify the common misconceptions they may possess that are related to the topic of study. These items are designed as quick checks for understanding and allow each student one attempt at each question. You can use the data collected to decide whether to assign additional resources to the class, or determine what individual or groups of students may need reinforcement or accelerated learning, prior to completing the Explore portion of the lesson.

> **TEACHER NOTE** Use this student response to evaluate students' prior knowledge of the basic principle of evolution and students' ability to explain observations in terms of the mechanisms of evolution. The Model Lesson provides information on common student misconceptions.

TEI Giraffe Neck

Before You Begin

What Do I already know about mechanisms for evolution?

> **TEACHER NOTE** This formative assessment item is intended to provide the teacher with feedback on how much students already understand about the role of natural selection and genetics in the process of evolutionary change.

- Image: Reach Up
- Image: Giraffe Necks

> **TEACHER NOTE** This item provides a formative pre-assessment of students' existing ability to distinguish between evolution and related concepts. Invite students to complete the item using think-pair-share. Then, challenge students to write and share their own examples for each concept.

TEI Evolution Vocabulary

> **TEACHER NOTE** Use this item to pre-assess students understanding of the connection between variation, genetic traits and adaptations.

TEI Success of Not

Explore `315 minutes`

Lesson Questions (LQs):

1. What are Charles Darwin's contributions to our understanding about evolution?
2. How does natural selection work?
3. How do natural selection and adaptation impact biodiversity?
4. How can the changes in allele frequency within a population be measured?
5. How does speciation occur?
6. What role do genetic drift, bottleneck effect, and the founder effect play in evolution?

Effective science instruction involves a student-centered rather than a teacher-centered approach. This can be accomplished either with Directed Inquiry or Guided Inquiry, depending on the needs and abilities of your class. Encourage students to select a variety of resources in their pursuit of answers as they work through Explore, with the end goal of constructing their scientific explanation in the Explain tab.

Directed Inquiry	Guided Inquiry
In Directed Inquiry, teachers provide students with a sequence of specified resources, challenging questions, and clear outcomes. Within this context students are given the opportunity to interact independently with each resource as prescribed by the teacher. Often different student groups can be guided through several different resources at the same time. For example, one group could work on a reading passage while a second group conducts a small- group Hands-On Activity with the teacher, and a third group is independently engaged with an online interactive resource.	In Guided Inquiry, students have independence to decide the scope and sequence of their investigations. Using resources from Tech book, students determine for themselves which resources they will Explore to answer the Lesson Questions. It is important to note that each student will choose multiple resources, but no one student is expected to use all the resources available. Students also determine the order in which to explore these resources and how to record their findings.

NGSS Components

SEP	CCC
■ Analyzing and Interpreting Data ■ Developing and Using Models ■ Using Mathematics and Computational Thinking ■ Constructing Explanations and Designing Solutions ■ Scientific Knowledge is based on Empirical Evidence ■ Using Mathematics and Computational Thinking	■ Cause and Effect ■ Systems and System Models ■ Scale, Proportion and Quantity ■ Patterns

Lesson Question: What are Charles Darwin's Contributions to Our Understanding About Evolution?

`Recommended 45 minutes`

TEACHER NOTE Connections: Crosscutting Concepts: Systems and System Models: In this concept, students will understand how evolution changes life on Earth by acting on individual organisms to bring about changes in the genetic makeup of populations. They use models to simulate changes produced by natural selection on populations

As students read and comprehend complex texts, view the videos, and complete the interactives, labs, and other Hands-On Activities, have them summarize and obtain scientific and technical information. Students will use this evidence to support their initial ideas on how to answer the Explain Question or their own question they generated during Engage. Have students record their evidence using "My Notebook."

- Core Interactive Text: What are Charles Darwin's contributions to our understanding about evolution?
- Video: A Brief Biography of Charles Darwin
- Video: Who Was Charles Darwin?
- Activity: Darwin and Wallace

Formative Assessment:

Throughout Explore, Technology Enhanced Items (TEIs) are embedded as multi-dimensional formative checks for understanding. You can use the data they provide to

- assign additional support
- extend learning
- design additional learning tasks to clarify student misconceptions

The Explore TEIs provide students with three attempts to demonstrate their proficiency. Scaffolded feedback is provided for each attempt. If a student does not achieve proficiency by the third attempt, a media asset is provided as an additional learning opportunity.

TEI Charles Darwin

Lesson Question: How Does Natural Selection Work?

`Recommended 45 minutes`

- Core Interactive Text: How does Natural Selection Work?
- Video: Traditional Plant Breeding
- Video: Why Are There So Many Breeds of Dogs?
- Video: The Origins of Modern Crops

TEACHER NOTE Misconception: Students may think evolution is a random process. In fact, evolution is usually driven by the process of natural selection, which is a nonrandom process.

- Image: Natural Selection in Action
- Video: The Evolution of Complex Organs
- Exploration: Evolution

TEI Natural or Artificial

Lesson Question: How Do Natural Selection and Adaptation Impact Biodiversity?

Recommended 45 minutes

- Core Interactive Text: How do Natural Selection and Adaptation Impact Biodiversity?

TEACHER NOTE Misconception: Students may think that evolution occurs in a relatively short amount of time. In fact, evolution usually takes place over many years, with evolution of new species often taking place over hundreds of thousands or millions of years. However, evolutionary processes are readily observable in existing populations.

- Image: Natural Selection: Examples from the Galapagos

TEACHER NOTE Practices: Science and Engineering Practice: Developing and Using Models: In this item, students must apply their understanding to a simple model of the relationship between traits, natural selection and changes in the frequency of a trait within a population. Have students create a similar scenario of their own in which a selective disadvantage would affect the frequency of a gene within a population.

TEI Birds

- Video: Types of Natural Selection

TEACHER NOTE Practices: Science and Engineering Practice: Mathematics and Computational Thinking: In this item, students use a mathematical model to support explanations, predict phenomena, and analyze systems. Help them summarize the information in the graphs by using the Six Word Story strategy. Have them predict what could happen to a population before and after an event to understand what caused the change in the allele frequency as shown in the graphs. The Six Word Story strategy is found on the Professional Learning tab. Click on Strategies & Resources, then click on Spotlight on Strategies (SOS). Now click on Summarizing, then click on Spotlight on Strategies: Six Word Story.

TEI Selection

Lesson Question: How Can the Changes in Allele Frequency within a Population Be Measured?

Recommended 90 minutes

- Core Interactive Text: How Can the Changes in Allele Frequency within a Population Be Measured?

TEACHER NOTE Practices: Science and Engineering Practice: Mathematics and Computational Thinking: The derivation and use of the Hardy-Weinberg Equilibrium is beyond the scope of what many students are expected to know at this grade level. However, if students are familiar with simple probabilities, a simple justification can be provided. If the students treat the gene pool like any population, you can say that for a gene with only two alleles, the probability of choosing any one of those genes is p and q. Then, ask the students, what is the probability of choosing two copies of allele p? If they are familiar with probability, they know that the probability would be the product of the individual probabilities, which is $p2$. This is a homozygote. It is similar to the probability of flipping a coin and obtaining heads twice in a row; it is the product of the probability of obtaining heads on each flip.

- Video: Basic Unit of Evolution
- Hand's-On Activity: Investigating the Hardy Weinberg Equilibrium

TEACHER NOTE Science and Engineering Practices: Analyzing and Interpreting Data: In this item, students use a mathematical model to understand population allele frequencies at equilibrium. They use a mathematical model to support explanations, predict phenomena, and analyze systems. Some students may be uncomfortable with the abstract variables used in the mathematical formulas for Hardy-Weinberg Equilibrium. Help them conceptualize the model in physical terms using a probability demonstration. Place 100 red balls and 100 blue balls in a bag. Have each student pick balls from the bag. Students with two colors of the same ball are homozygotes. Students with two different colored balls are heterozygotes. Ask them how closely their physical demonstration aligns with the predictions of the abstract model.

TEI Equilibrium

Lesson Question: How Does Speciation Occur?

Recommended 45 minutes

- Core Interactive Text: How Does Speciation Occur?
- Video: Speciation
- Image: Sexual Dimorphism
- Video: Sexual Selection

Lesson Question: What Role Do Genetic Drift, Bottleneck Effect, and the Founder Effect Play in Evolution?

Recommended 45 minutes

- Core Interactive Text: What Role Do Genetic Drift, Bottleneck Effect, and the Founder Effect Play in Evolution?
- Video: Genetic Drift

TEACHER NOTE Connections: Crosscutting Concepts: Cause and Effect: In this item, students understand the evolutionary effects of different pressures that drive evolution. They suggest cause-and-effect relationships to explain and predict behaviors in complex natural systems. These concepts require students to learn many new scientific vocabulary words. Help them learn these new terms using the Vocabulary Scavenger Hunt strategy. As they watch the videos and read the Core Interactive Text, have them note whenever a new term appears. The "Vocabulary Scavenger Hunt" strategy is found on the Professional Learning tab. Click on Strategies & Resources, then click on Spotlight on Strategies (SOS). Now click on Vocabulary Development, then click on Spotlight on Strategies: Vocabulary Scavenger Hunt.

TEI Sources of Evolution

Explore More Resources

Resources in Explore More Resources support differentiation within your classroom by

- providing additional visualization of content
- affording extension of content to those students ready for acceleration
- offering Lexile reading levels for reading passages

Online explorations and hands-on experiences are provided so that students can conduct virtual investigations, collect and design investigations, and collect and analyze data; these skills are essential to developing scientific understanding.

Explain `45–90 minutes`

In Explore, students
1. uncovered scientific understandings
2. conducted investigations
3. analyzed data, text, and other media resources
4. collected evidence to support their scientific explanation

In Explain, provide students with time to formally compose their scientific explanations around the Explain or student-generated questions using evidence collected from Explore.

Scientific explanations are student responses, either written or orally presented, that explain scientific phenomena based upon evidence. Developing a scientific explanation requires students to analyze and interpret data to construct meaning out of the data. There are three main components to the scientific explanation: the claim, the evidence, and the reasoning.

To help students to communicate their scientific explanations, allow them to utilize the multimedia creation tools such as Studio and Whiteboard. Remind them that they may upload image, audio, and video files using the "attach file" option to communicate their scientific explanations.

Students may construct their scientific explanations individually or within a small group of students. Students should communicate their explanations with other classmates, and provide constructive criticism and refine their explanations prior to submission to the teacher. If explanations are used as a formative assessment, you can provide additional feedback and comments to support students as they refine their explanations.

EXPLAIN
How does evolution work to create the diversity of life on Earth?

Elaborate with STEM `45–135 minutes`

*Elaborate with STEM are optional extension resources available after students have demonstrated proficiency with standards addressed previously in the concept.

NGSS Components

SEP	CCC
■ Analyzing and Interpreting Data ■ Obtaining, Evaluating, and Communicating Information ■ Planning and Carrying Out Investigations	■ Cause and Effect ■ Structure and Function

STEM In Action `45 minutes`

STEM in Action ties the scientific concepts to real-world applications, with many connecting to STEM careers. Technology Enhanced Items (TEIs) expect students to critically read the Core Interactive Text (CIT) and review the provided media resources.

Applying Mechanisms for Evolution

- Core Interactive Text: Applying Mechanisms for Evolution
- Video: Resistance to Antibiotics
- Image: Antibiotic-Free Milk
- Video: Evolution and Technology
- Image: The Invention of the Hook and Loop Fastener

TEI Antibiotic Resistance

STEM Project Starters

STEM Project Starters provide additional real-world contexts that require students to apply and extend their content knowledge related to the concept. STEM Project Starters can also serve as an alternative instructional hook presented at the beginning of the learning progression. The project can then be revisited throughout and at the end of the 5E learning cycle, for students to apply content knowledge.

STEM Project Starter: The Timeline of Evolution

Recommended 45 minutes

Why have humans changed over time? What factors led to these changes and why did they happen?

- Image: The Evolution of Humans

TEACHER NOTE This project can be used as either a formative or summative assessment. It is designed to emphasize the fact that evolutionary change takes time and that with each change, there has to be some environmental influence. Assign this project to individual students, as each will have a different preference as to which product they want to investigate. It is okay if two students select the same product from which they build their models.

TEI The Timeline of Evolution

STEM Project Starter: Natural Selection and the Peppered Moths

Recommended 90 minutes

The peppered moth of England is often used as a model to show natural selection in action. Why is this animal such a great model?

- Image: Peppered Moths Show Natural Selection

TEACHER NOTE This can be used as a summative assessment to test students' knowledge of natural selection. Have students work in teams to research and design their experiments.

It should be completed after students have done the simulation "Evolution" to use as reinforcement about how species change as a result of their environment. By doing this online activity first, students will have a much better idea of what needs to be done when designing their experiments about the peppered moth.

- Exploration: Evolution

TEI Natural Selection Experiment

UNIT 3: Evolving Earth

Evaluate `45–90 minutes`

Explain Question:
How does evolution work to create the diversity of life on Earth?

Lesson Questions (LQ):
- What are Charles Darwin's contributions to our understanding about evolution?
- How does natural selection work?
- How do natural selection and adaptation impact biodiversity?
- How can the changes in allele frequency within a population be measured?
- How does speciation occur?
- What role do genetic drift, bottleneck effect, and the founder effect play in evolution?

Throughout instruction and the 5E learning cycle, you will have collected formative assessment data to drive the assignment of resources and experiences to students. Evaluate is intended to include summative assessment checks for proficiency. You can use the Explain and Lesson Questions for the concept as a summative assessment in a variety of ways such as these:

- Post each Lesson Question (LQ) in various locations in the classroom, and have small groups of students generate claim statements related to the Lesson Question (LQ). Other students can add to the claim, or refute the claim, during a gallery walk where they place additional pieces of evidence on each Lesson Question (LQ) poster.
- Assign small groups of students to each Lesson Question (LQ) and have the groups generate a poster, board, graphic, or piece of text that answers the question. Use a jigsaw approach and create a second set of groups that contain members from each Lesson Question (LQ) group to share their ideas.
- Ask students to return to their initial ideas for the Explain question and add additional details and evidence.

Encourage students to review the concept review and complete the Student Self-Check practice assessment prior to assigning the Summative Teacher Concept assessment.

- Student Review and Practice Assessment
- Teacher Concept Assessments

PROBLEM SETS

Calculating a Hardy-Weinberg Equation

Use the Hardy-Weinberg equation to solve the problems below.

1. Suppose the frequency of the dominant allele in a population is 75%. What is the frequency of the recessive allele?
   answer: 25%

2. A population of salamanders has two colors, blue and white. Blue is the dominant color and white is recessive. The frequency of the white allele is 36%. What percentage of the salamander population is white?
   answer: 13% of the salamander population is white.

3. A population of 100 individuals has 49 individuals that are homozygous dominant for a certain trait. How many individuals are heterozygous for the trait? How many are homozygous recessive for the trait?
   answer: Rr = 42; rr = 9

4. The frequency of a dominant allele (P) in a population is 60% and the frequency of the recessive allele (p) is 40%. Predict the probability of the possible genotypes in this population.
   answer: PP = 36%; Pp = 48%; pp = 16%

5. A population has two alleles A and a for a gene, with phenotypic frequencies: 20:60:20. Is this population in Hardy-Weinberg equilibrium? Explain your answer
   answer: Students will need to calculate the observed genotype and allele frequencies. The allele frequencies are then used to predict genotype frequencies. The predicted genotype frequencies are compared with the observed genotype frequencies. If the difference is significant, the population is not in Hardy-Weinberg equilibrium.

   Observed genotype and allele frequencies
   Total population = 20 + 60 + 20 = 100
   AA = 20/100 = 0.2
   Aa = 60/100 = 0.6
   aa = 20/100 = 0.2

   A = (40+60)/200 = 100/200 = 0.5
   a = (40+60)/200 = 100/200 = 0.5

   Predicted genotype frequencies
   p^2 = 0.5 × 0.5 = 0.25
   2pq = 2(0.5 × 0.5) = 0.5
   q^2 = 0.5 × 0.5 = 0.25

   The difference between 0.25:0.5:0.25 and 0.2:0.6:0.2 indicates that the population is not in Hardy-Weinberg equilibrium.

Genetics

The Five E Instructional Model

Science Techbook follows the 5E instructional model. This Model Lesson includes strategies for each of the 5Es. As you design the inquiry-based learning experience for students, be sure to collect data during instruction to drive your instructional decisions. Point-of-use teacher notes are also provided within each E-tab.

Engage 45–90 minutes

Engage Media Resources

The resources found in Engage are intended to stimulate students by exposing them to a phenomenon relevant to the content of the lesson. Engage also provides examples of relevant real-world applications that allow students to begin to make observations and relate the science content to their everyday lives. The Core Interactive Text (CIT) and media resources are carefully designed to prompt students to begin asking questions that they can investigate during the Explore phase of the lesson. They should also start collecting evidence to address the Explain question located at the bottom of the Engage page.

> **TEACHER NOTE** **Investigative Phenomenon:** In Engage, students are stimulated to think about where their different and similar characteristics come from. We are all human, but we are also all easily recognizable as individuals. We have many characteristics that define us as human—two arms, two legs, a large brain, etc.—but many characteristics that are different—hair color, skin shade, height, and more. How does this apparent contradiction between similarity and variation come about? Students will have encountered the basic mechanism behind inheritance in earlier grades so should be able to reference the idea that different genes have different alleles. Use the video as a catalyst to initiate group conversations about what they already know about inheritance. Have each group generate a list of what they know and want to know about inheritance and variation. Use this to create a consolidated class list on flipchart paper and refer to it when their questions are addressed in this unit.

- Core Interactive Text: Similar but Different
- Video: Our Characteristics
- Image: Chromosomes
- Video: Heredity: How Our Parents' Genes Affect Us
- Interactive Video: Heredity

Explain Question

The Explain question focuses students on gathering information in the Explore section. The Explain question can be used to

- Record what students already know related to the Explain question.
- Serve as a template or model for students to generate their own scientific questions.
- Collect evidence as students work through the lesson.
- Allow students to reflect on their growth before and after the lesson.

Explain

How does the genetic information inherited from an organism's parents affect its characteristics?

- Image: Eyes

Engage Formative Assessment

Technology Enhanced Items (TEIs) found on the Engage page enable you to collect data on students' prior knowledge and identify the common misconceptions they may possess that are related to the topic of study. These items are designed as quick checks for understanding and allow each student one attempt at each question. You can use the data collected to decide whether to assign additional resources to the class, or determine what individual or groups of students may need reinforcement or accelerated learning, prior to completing the Explore portion of the lesson.

TEACHER NOTE Use this student response as a formative assessment to evaluate students' knowledge of genetics, and to make a connection between traits other than physical appearance and herecity. Suggested use of this activity includes teacher-led whole-group discussion of their thought about how knowledge of family health can help make lifestyle decisions.

TEI Genetics and Adoption

Before You Begin

What Do I Already Know about Genetics?

TEACHER NOTE This preconception activity (preassessment) is intended as a formative assessment to help guide instruction and to stimulate student thinking and assess prior knowledge. Suggested use of the item with students involves a whole-class or small-group discussion, in which each term is discussed by students to help formulate their answers.

TEI Genetic Material

TEACHER NOTE This item identifies misconceptions about genetics, particularly the notion that recessive genes are inferior to dominant ones. Students have likely already been introduced to the concept of recessive and dominant genes. Use this formative assessment item to stimulate class discussion about the content of this lesson.

TEI Describing the Connection

TEACHER NOTE This preconception activity (preassessment) is intended as a formative assessment to help guide instruction and to stimulate student thinking and assess prior knowledge. Suggested use of the item with students involves a whole-class or small-group discussion in which each statement is discussed by students to help formulate their answers.

TEI DNA and Reproduction

TEACHER NOTE This item identifies misconceptions about genetics, particularly the notion that recessive genes are inferior to dominant ones. Students have likely already been introduced to the concept of recessive and dominant genes. Use this formative assessment item to stimulate class discussion about the content of this lesson.

TEI Understanding Genetics

- Video: Parent Genes and Offspring
- Video: The Human Genome Project
- Video: Alleles, Genes, and Chromosomes

Explore `225 minutes`

Lesson Questions (LQs):

1. What is Mendelian inheritance?
2. What is the difference between dominant and recessive alleles?
3. What are the definitions of homozygous, heterozygous, genotype, and phenotype?
4. What are Mendel's laws of inheritance?
5. How are the results of monohybrid and dihybrid crosses diagrammed?
6. What are the effects of multiple alleles, codominance, and incomplete dominance on phenotype?

Effective science instruction involves a student-centered rather than a teacher-centered approach. This can be accomplished either with Directed Inquiry or Guided Inquiry, depending on the needs and abilities of your class. Encourage students to select a variety of resources in their pursuit of answers as they work through Explore, with the end goal of constructing their scientific explanation in the Explain tab.

Directed Inquiry	Guided Inquiry
In Directed Inquiry, teachers provide students with a sequence of specified resources, challenging questions, and clear outcomes. Within this context students are given the opportunity to interact independently with each resource as prescribed by the teacher. Often different students groups can be guided through several different resources at the same time. For example, one group could work on a reading passage while a second group conducts a small-group Hands-On Activity with the teacher, and a third group is independently engaged with an online interactive resource.	In Guided Inquiry, students have independence to decide the scope and sequence of their investigations. Using resources from Techbook, students determine for themselves which resources they will Explore to answer the Lesson Questions. It is important to note that each student will choose multiple resources, but no one student is expected to use all the resources available. Students also determine the order in which to explore these resources and how to record their findings.

NGSS Components

SEP	CCC
■ Engaging in Argument from Evidence ■ Obtaining, Evaluating, and Communicating Information	■ Patterns ■ Cause and Effect

Lesson Question: What Is Mendelian Inheritance?

Recommended 45 minutes

- Core Interactive Text: What Is Mendelian Inheritance?

TEACHER NOTE Practices: Science and Engineering Practice: Engaging in Argument from Evidence: Throughout this lesson, students explore the historical contributions that Gregor Mendel made to our understandings of genetic inheritance. Mendel's experimental design modeled the need to compare and evaluate claims based on evidence and reasoning. Using the data from his numerous experiments with pea plants, Mendel constructed the laws of segregation and of independent assortment that became part of the foundation for modern genetics. As an extension, have students brainstorm observed human characteristics and traits. Emphasize that every characteristic can have more than one trait. As students list the characteristics and traits, respectfully encourage critiques of each classification. For example, students might say hair color is a trait or having blue eyes is a characteristic. After creating the list, have students construct a claim of the relationship between characteristics and traits. Encourage students to use evidence to support their claims.

- Image: Plant Traits
- Video: Patterns of Inheritance: Mendel's Contribution to Genetics

Formative Assessment:

Throughout Explore, Technology Enhanced Items (TEIs) are embedded as multi-dimensional formative checks for understanding. You can use the data they provide to

- assign additional support
- extend learning
- design additional learning tasks to clarify student misconceptions

The Explore TEIs provide students with three attempts to demonstrate their proficiency. Scaffolded feedback is provided for each attempt. If a student does not achieve proficiency by the third attempt, a media asset is provided as an additional learning opportunity.

TEACHER NOTE Connections: Crosscutting Concept: Cause and Effect: In the following item, students will select steps that describe elements of Gregor Mendel's scientific process and findings. Mendel conducted scientific investigations, breeding different variations in pea plants, in order to find the underlying cause of traits. Before students work through the item, emphasize Mendel's need to use empirical evidence to differentiate between cause and correlation. Students will propose a relationship between the cause of traits and the effect of traits on the larger natural system. Ask: What does it mean for a plant to be true-breeding? Then, students should work with the rest of the statements to show how Mendel determined cause and effect relationships, explaining and predicting behaviors in natural systems.

TEI Mendelian Inheritance

Lesson Question: What Is the Difference between Dominant and Recessive Alleles?

`Recommended 20 minutes`

> **TEACHER NOTE** **Misconception:** Students may associate "recessive" with inferiority or a less adaptive genotype. Explain that dominant genotypes are not superior to recessive genotypes. There is no genotype that is better than another; each genotype is simply a different arrangement of alleles.

- Core Interactive Text: What Is the Difference between Dominant and Recessive Alleles?
- Video: Complete Dominance

Lesson Question: What Are the Definitions of Homozygous, Heterozygous, Genotype, and Phenotype?

`Recommended 35 minutes`

- Core Interactive Text: What Are the Definitions of Homozygous, Heterozygous, Genotype, and Phenotype?
- Video: Particulate Inheritance
- Image: Heterozygous
- Image: Homozygous and Heterozygous
- Video: Genotype and Phenotype
- Image: Horned Lizard
- Image: Genotypes and Phenotypes

> **TEACHER NOTE** **Connection: Crosscutting Concept: Patterns:** A fundamental understanding of the distinction between phenotype and genotype is necessary before students will understand how to calculate the phenotypic and genotypic ratios derived from a Punnett square. This item will assess student understanding of the patterns involved determining if pairs of alleles are homozygous or heterozygous and how the pattern affects the expression of dominant and recessive traits.
>
> Before proceeding to the item, you may wish to use a "Stations" activity. For this activity, choose one trait with multiple variations, such as eye color. Set up several stations with images of different expressions of that trait (for example, an image of green eyes at one table, blue eyes at another, brown at another). Provide students at each station with a Punnett square of alleles from two parents that could have produced the given eye color. For example, BB and bb, signifying one parent with two brown alleles and the other with two blue alleles. Have students use the patterns involved in the transmission of alleles to identify how many possible genotypes on the Punnett square could lead to the given outcome, and then express this outcome as a ratio. Students should rotate tables until they have completed all stations. Have students compare the mathematical ratios for each station to determine resulting offspring patterns for the following parents: two homozygous dominant, two homozygous recessive, two heterozygous, one homozygous dominant and one heterozygous, and one homozygous recessive and one heterozygous.

TEI Forms and Expressions of Alleles

UNIT 4: Inheritance and Variation

Lesson Question: What Are Mendel's Laws of Inheritance? **Recommended 25 minutes**

- Core Interactive Text: What Are Mendel's Laws of Inheritance?
- Video: Mendel's Laws of Inheritance
- Image: Traits in Peapods
- Video: Types of Reproduction

Lesson Question: How Are the Results of Monohybrid and Dihybrid Crosses Diagrammed? **Recommended 50 minutes**

- Core Interactive Text: How Are the Results of Monohybrid and Dihybrid Crosses Diagrammed?
- Image: Punnett Square
- Video: Using the Punnett Square
- Video: Punnett Square
- Exploration: Genetics
- Video: Monohybrid vs. Dihybrid Cross

TEACHER NOTE Practices: Science and Engineering Practice: Obtaining, Evaluating, and Communicating Information: A PDF version of this item is available in Explore More Resources. This item allows students to apply their understanding of Punnett squares to compute the probability of the two phenotypic and genotypic ratios. You may wish to use an introductory strategy with your class in which you use an image of a model selected beforehand in a "Wheel of Fortune" type game.

Select one characteristic with several traits to work with, such as hair (color and texture), and assign each allele a letter (for example, color could be *B*, brown, or *b*, blonde, while texture could be *C*, curly, or *c*, straight). Create a dihybrid cross Punnett square that applies to the image you have selected and post it on a board or projection before class, but cover the text of each square.

Students will then take turns suggesting possible genotypes that could lead to the expressed phenotype, until the entire Punnett square has been uncovered.

TEI Homozygous Dominant

TEI First Offspring

TEI First Two Offspring

4.1 Genetics

Lesson Question: What Are the Effects of Multiple Alleles, Codominance, and Incomplete Dominance on Phenotype?

Recommended 50 minutes

- Core Interactive Text: What Are the Effects of Multiple Alleles, Codominance, and Incomplete Dominance on Phenotype?
- Video: Blood Groups Result from Multiple Alleles
- Image: Blood Types
- Image: Roan Steer
- Video: Comparing Complete Dominance and Codominance
- Video: What Is Codominance?
- Image: Incomplete Dominance
- Video: How Does Incomplete Dominance Work?
- Video: Sex Determining Chromosomes

TEACHER NOTE Practices: Science and Engineering Practice: Obtaining, Evaluating, and Communicating Information: This item allows students to apply their understanding of Mendel's two laws— the law of segregation and the law of independent assortment—to assess statements that contradict these laws. Have students critically read each statement to evaluate its validity. It may be helpful for students to set up Punnett squares to compare if the statement matches the evidence. Extend this item by having students use the Internet to gather and read examples of Mendel's laws or possible mistakes by modern media which contradict these laws. Have students verify the correct examples using data from class or other valid sources. Have students communicate why some of the examples are incorrect and contradict Mendel's laws. This extension can be an ongoing project, giving students the opportunity to research, organize, evaluate, and communicate information about inheritance.

TEI Laws of Inheritance

Explore More Resources

Resources in Explore More Resources support differentiation within your classroom by

- providing additional visualization of content
- affording extension of content to those students ready for acceleration
- offering Lexile reading levels for reading passages

Online explorations and hands-on experiences are provided so that students can conduct virtual investigations, collect and design investigations, and collect and analyze data; these skills are essential to developing scientific understanding.

Explain `45–90 minutes`

In Explore, students
1. uncovered scientific understandings
2. conducted investigations
3. analyzed data, text, and other media resources
4. collected evidence to support their scientific explanation

In Explain, provide students with time to formally compose their scientific explanations around the Explain or student-generated questions using evidence collected from Explore.

Scientific explanations are student responses, either written or orally presented, that explain scientific phenomena based upon evidence. Developing a scientific explanation requires students to analyze and interpret data to construct meaning out of the data. There are three main components to the scientific explanation: the claim, the evidence, and the reasoning.

To help students to communicate their scientific explanations, allow them to utilize the multimedia creation tools such as Board Builder and Whiteboard. Remind them that they may upload image, audio, and video files using the "attach file" option to communicate their scientific explanations.

Students may construct their scientific explanations individually or within a small group of students. Students should communicate their explanations with other classmates, and provide constructive criticism and refine their explanations prior to submission to the teacher. If explanations are used as a formative assessment, you can provide additional feedback and comments to support students as they refine their explanations.

EXPLAIN

How does the genetic information inherited from an organism's parents affect its characteristics?

Elaborate with STEM `45–135 minutes`

*Elaborate with STEM are optional extension resources available after students have demonstrated proficiency with standards addressed previously in the concept.

NGSS Components

SEP	CCC
■ Asking Questions and Defining Problems ■ Planning and Carrying Out Investigations ■ Constructing Explanations and Designing Solutions ■ Obtaining, Evaluating, and Communicating Information	■ Patterns ■ Cause and Effect ■ Structure and Function

STEM In Action `45 minutes`

STEM in Action ties the scientific concepts to real-world applications, with many connecting to STEM careers. Technology Enhanced Items (TEIs) expect students to critically read the Core Interactive Text (CIT) and review the provided media resources.

Applying Genetics

- Core Interactive Text: Applying Genetics
- Image: Genetically Engineered Algae
- Video: Algae Biofuel
- Video: Spider Silk from Goats
- Video: RNA Interference
- Video: What Are Restriction Enzymes?
- Video: How Do Restriction Enzymes Work?
- Video: Uses of Restriction Enzymes

TEACHER NOTE Use this student response as a formative assessment to evaluate students' knowledge of the genetic traits and to think about how they can be useful. Suggested use of this item includes students working in think-pair-shares to discuss their thoughts with a partner.

- Image: Genetically Engineered Trees

TEI Creating a Forest

STEM Project Starters

STEM Project Starters provide additional real-world contexts that require students to apply and extend their content knowledge related to the concept. STEM Project Starters can also serve as an alternative instructional hook presented at the beginning of the learning progression. The project can then be revisited throughout and at the end of the 5E learning cycle, for students to apply content knowledge.

STEM Project Starter: Solve the Pea Problem   `Recommended 45 minutes`

How can I grow only green peas?

> **TEACHER NOTE** Use this project as a formative assessment. Students can work in small groups to come up with a design for the process, but each student should submit their own work.

■ Exploration: Genetics

TEI Solve the Pea Problem

STEM Project Starter Frankenfood Alert   `Recommended 90 minutes`

Why are GMOs sometimes called "frankenfood"?

> **TEACHER NOTE** Students will have two submissions for this project. Ask them to turn in an outline of their research before developing the public service announcement. Evaluate the research for completeness and accuracy. Use as a formative assessment.
>
> Discuss with students how to judge the reliability of websites.

TEI The GMOs Case

Engage `45–90 minutes`

Explain Question:
How does the genetic information inherited from an organism's parents affect its characteristics?

Lesson Questions (LQ):
- What is Mendelian inheritance?
- What is the difference between dominant and recessive alleles?
- What are the definitions of homozygous, heterozygous, genotype, and phenotype?
- What are Mendel's laws of inheritance?
- How are the results of monohybrid and dihybrid crosses diagrammed?
- What are the effects of multiple alleles, codominance, and incomplete dominance on phenotype?

Throughout instruction and the 5E learning cycle, you will have collected formative assessment data to drive the assignment of resources and experiences to students. Evaluate is intended to include summative assessment checks for proficiency. You can use the Explain and Lesson Questions for the concept as a summative assessment in a variety of ways such as these:

- Post each Lesson Question (LQ) in various locations in the classroom, and have small groups of students generate claim statements related to the Lesson Question (LQ). Other students can add to the claim, or refute the claim, during a gallery walk where they place additional pieces of evidence on each Lesson Question (LQ) poster.
- Assign small groups of students to each Lesson Question (LQ) and have the groups generate a poster, board, graphic, or piece of text that answers the question. Use a jigsaw approach and create a second set of groups that contain members from each Lesson Question (LQ) group to share their ideas.
- Ask students to return to their initial ideas for the Explain question and add additional details and evidence.

Encourage students to review the concept review and complete the Student Self-Check practice assessment prior to assigning the Summative Teacher Concept assessment.

- Student Review and Practice Assessment
- Teacher Concept Assessments

PROBLEM SETS

Dihybrid Cross

Use dihybrid cross to solve the problems below.

Dihybrid Cross for Hair Type and Eye Color

Genotype	HA	Ha	hA	ha
HA	HHAA	HHAa	HhAA	HhAa
Ha	HHAa	HHaa	HhAa	Hhaa
hA	HhAA	HhAa	hhAA	hhAa
ha	HhAa	Hhaa	hhAa	hhaa

H = long hair
h = short hair
A = brown
a = green

1. What is the probability of having offspring that are homozygous dominant for both traits?

    answer: $\frac{1}{16}$ or 6%

2. What is the probability that the first offspring will be either long-haired and brown-eyed or long-haired and green-eyed?

    answer: $(\frac{9}{16}) + (\frac{3}{16}) = (\frac{12}{16})$ or 75%

3. What is the probability that among the first two offspring, one will be short-haired and green-eyed and the other will be short-haired and brown-eyed?

    answer: $(\frac{1}{16}) \times (\frac{3}{16}) = (\frac{3}{256})$ or 1%

DNA

The Five E Instructional Model

Science Techbook follows the 5E instructional model. This Model Lesson includes strategies for each of the 5Es. As you design the inquiry-based learning experience for students, be sure to collect data during instruction to drive your instructional decisions. Point-of-use teacher notes are also provided within each E-tab.

Engage   45–90 minutes

Engage Media Resources

The resources found in Engage are intended to stimulate students by exposing them to a phenomenon relevant to the content of the lesson. Engage also provides examples of relevant real-world applications that allow students to begin to make observations and relate the science content to their everyday lives. The Core Interactive Text (CIT) and media resources are carefully designed to prompt students to begin asking questions that they can investigate during the Explore phase of the lesson. They should also start collecting evidence to address the Explain question located at the bottom of the Engage page.

> **TEACHER NOTE** **Investigative Phenomenon:** The phenomenon of genetically modified glowing aquarium fish is used to activate students' prior knowledge about the role of DNA in coding for traits. The fact that scientists transferred a trait from one unrelated organism to another provides a clear and easy to follow example as to how a gene consisting of a specific sequence of DNA codes for a specific trait. The example also provides a connection between a gene and the synthesis of a protein, which is an important introduction to the way DNA codes (via RNA) for the specific sequence of amino acids in a protein. Extend and engage by having students research other examples of genetically modified organisms, and the function of the proteins that the genes that were introduced into their genome code for.

- Core Interactive Text: Glowing Fish
- Image: Glowing Fish
- Video: The Zebrafish

Explain Question

The Explain question focuses students on gathering information in the Explore section. The Explain question can be used to

- Record what students already know related to the Explain question.
- Serve as a template or model for students to generate their own scientific questions.
- Collect evidence as students work through the lesson.
- Allow students to reflect on their growth before and after the lesson.

Explain

How does DNA carry the code for a specific protein?

- Image: DNA Structure

Engage Formative Assessment

Technology Enhanced Items (TEIs) found on the Engage page enable you to collect data on students' prior knowledge and identify the common misconceptions they may possess that are related to the topic of study. These items are designed as quick checks for understanding and allow each student one attempt at each question. You can use the data collected to decide whether to assign additional resources to the class, or determine what individual or groups of students may need reinforcement or accelerated learning, prior to completing the Explore portion of the lesson.

TEACHER NOTE Use this student response to evaluate students' prior knowledge of the concept. The Model Lesson provides information on common student misconceptions.

TEI DNA, RNA, and Proteins

Before You Begin

What Do I Already Know about DNA?

TEACHER NOTE This formative assessment item is intended to provide the teacher with feedback on prior knowledge of this topic. In middle school, students should have learned the differences between atoms and compounds and should have an understanding of what these are and how they are related. This would be good to use as an introductory activity.

TEI Atoms and Molecules

TEACHER NOTE This is a formative assessment item. Students selecting A are not aware that enzymes do not have any coloration. Students selecting C probably have the misconception that enzymes are able to function wherever they are located as long as other factors are sufficient. This question would be effective when answered individually.

TEI Enzymes and DNA

TEACHER NOTE This is a formative assessment item that would be effective when answered individually. Students should recognize the difference between a learned behavior and a physical trait or characteristic. If students select anything other than A, discuss with them how learned behaviors are not inheritable, while some behavioral traits may be heritable. This may help identify existing misconceptions.

TEI Inheritance

- Video: Atoms, Molecules, and Compounds
- Video: Discovery of DNA Structure
- Video: Proteins
- Reading Passage: Unlocking the Door to the Spiral Staircase

Explore `135 minutes`

Lesson Questions (LQs):

1. What are the components of DNA and RNA?
2. How are the structures of DNA and RNA related to their functions?
3. How do cells make copies of DNA?

Effective science instruction involves a student-centered rather than a teacher-centered approach. This can be accomplished either with Directed Inquiry or Guided Inquiry, depending on the needs and abilities of your class. Encourage students to select a variety of resources in their pursuit of answers as they work through Explore, with the end goal of constructing their scientific explanation in the Explain tab.

Directed Inquiry	Guided Inquiry
In Directed Inquiry, teachers provide students with a sequence of specified resources, challenging questions, and clear outcomes. Within this context students are given the opportunity to interact independently with each resource as prescribed by the teacher. Often different students groups can be guided through several different resources at the same time. For example, one group could work on a reading passage while a second group conducts a small-group Hands-On Activity with the teacher, and a third group is independently engaged with an online interactive resource.	In Guided Inquiry, students have independence to decide the scope and sequence of their investigations. Using resources from Techbook, students determine for themselves which resources they will Explore to answer the Lesson Questions. It is important to note that each student will choose multiple resources, but no one student is expected to use all the resources available. Students also determine the order in which to explore these resources and how to record their findings.

NGSS Components

SEP	CCC
■ Developing and Using Models ■ Constructing Explanations and Designing Solutions	■ Structure and Function

Lesson Question: What Are the Components of DNA and RNA? `Recommended 45 minutes`

- Core Interactive Text: What Are the Components of DNA and RNA?

TEACHER NOTE **Connections: Crosscutting Concept: Structure and Function:** In this concept, students investigate living systems by examining the properties and structures of different nucleic acids and how they contribute to life. They infer the functions and properties of DNA and the molecular substructures of nucleic acids. Engage students using the "Twenty Questions" strategy by displaying images of models of DNA and RNA having students pose questions. The Twenty Questions strategy is found on the Professional Learning tab. Click on Strategies & Resources, then Spotlight On Strategies (SOS). "Twenty Questions" is found underneath "Questioning."

- Video: Building Blocks of DNA
- Video: Paired DNA Strands
- Video: DNA Packaging
- Video: DNA in the Cell
- Video: Human Chromosomes

TEACHER NOTE **Misconception:** Students may think that each organism has a different kind of DNA. In fact, DNA is the same molecule, made of the same subunits, in all species—from bacteria to plants to humans. The only differences are the amount of DNA that may be present and the order of the base codes.

Formative Assessment:

Throughout Explore, Technology Enhanced Items (TEIs) are embedded as multi-dimensional formative checks for understanding. You can use the data they provide to

- assign additional support
- extend learning
- design additional learning tasks to clarify student misconceptions

The Explore TEIs provide students with three attempts to demonstrate their proficiency. Scaffolded feedback is provided for each attempt. If a student does not achieve proficiency by the third attempt, a media asset is provided as an additional learning opportunity.

TEACHER NOTE **Practices: Science and Engineering Practice: Developing and Using Models:** In this item, students use models of DNA and RNA to infer how they differ and to illustrate the relationship between components of the genetic code system. Help students see the differences between DNA and RNA by having them build or draw their own structures of DNA and RNA. Have them build or draw the models side by side. Every time the model is different in molecular structure, have them highlight it in a different color. Ask them to consider how these differences contribute to the functions of the different molecules.

TEI **DNA and RNA**

Lesson Question: How Are the Structures of DNA and RNA Related to Their Functions?

`Recommended 45 minutes`

TEACHER NOTE **Misconception:** Students may think that genes are separate from DNA. In fact, genes are segments of a DNA molecule that code for proteins.

- Core Interactive Text: How Are the Structures of DNA and RNA Related to Their Functions?
- Video: Triplet Code

TEACHER NOTE This concept introduces students to three related concepts: replication, transcription, and translation. Remind students that translation is different from replication and transcription because it involves the polymerization of amino acids, not nucleic acids. As a result, translation uses ribosomes rather than a DNA or RNA polymerase. Translation also occurs at ribosomes in the cytoplasm, while replication and transcription occur in the nucleus.

- Video: Transcription

Lesson Question: How Do Cells Make Copies of DNA?

`Recommended 45 minutes`

TEACHER NOTE **Misconception:** Students may think that one strand of DNA in each DNA molecule is inherited from each parent. In fact, the strands are complementary—they cannot code for different genes. Offspring of sexual reproduction inherit a set of chromosomes from each parent.

- Core Interactive Text: How Do Cells Make Copies of DNA?
- Video: Prokaryotic Cell Division
- Image: Lagging Strand
- Video: DNA Replication
- Video: Details of DNA Replication
- Exploration: DNA
- Hands-On Activity: Modeling DNA

TEACHER NOTE **Practices: Science and Engineering Practice: Constructing Explanations and Designing Solutions:** In this collection, students apply scientific ideas, principles, and evidence to provide an explanation of how DNA replication occurs in cells. Help them visualize the process of DNA replication with manipulatives in the classroom. Provide a model of double-stranded DNA and individual DNA and RNA nucleotides to each group of students. Have them show the unzipping of DNA and the polymerization of the leading and lagging strands away from the origin of replication. At the end of the exercise, students should see that the ends of linear DNA will result in 3' overhangs.

TEI DNA Replication

TEI Telomeres

Explore More Resources

Resources in Explore More Resources support differentiation within your classroom by

- providing additional visualization of content
- affording extension of content to those students ready for acceleration
- offering Lexile reading levels for reading passages

Online explorations and hands-on experiences are provided so that students can conduct virtual investigations, collect and design investigations, and collect and analyze data; these skills are essential to developing scientific understanding.

Explain `45–90 minutes`

In Explore, students
1. uncovered scientific understandings
2. conducted investigations
3. analyzed data, text, and other media resources
4. collected evidence to support their scientific explanation

In Explain, provide students with time to formally compose their scientific explanations around the Explain or student-generated questions using evidence collected from Explore.

Scientific explanations are student responses, either written or orally presented, that explain scientific phenomena based upon evidence. Developing a scientific explanation requires students to analyze and interpret data to construct meaning out of the data. There are three main components to the scientific explanation: the claim, the evidence, and the reasoning.

To help students to communicate their scientific explanations, allow them to utilize the multimedia creation tools such as Board Builder and Whiteboard. Remind them that they may upload image, audio, and video files using the "attach file" option to communicate their scientific explanations.

Students may construct their scientific explanations individually or within a small group of students. Students should communicate their explanations with other classmates, and provide constructive criticism and refine their explanations prior to submission to the teacher. If explanations are used as a formative assessment, you can provide additional feedback and comments to support students as they refine their explanations.

EXPLAIN
How does DNA carry the code for a specific protein?

Elaborate with STEM `45–135 minutes`

*Elaborate with STEM are optional extension resources available after students have demonstrated proficiency with standards addressed previously in the concept.

SEP	CCC
■ Asking Questions and Defining Problems ■ Developing and Using Models ■ Planning and Carrying Out Investigation ■ Using Mathematics and Computational Thinking ■ Constructing Explanations and Designing Solutions ■ Obtaining, Evaluating, and Communicating Information	■ Cause and Effect ■ Scale, Proportion, and Quantity ■ Structure and Function

STEM In Action `45 minutes`

STEM in Action ties the scientific concepts to real-world applications, with many connecting to STEM careers. Technology Enhanced Items (TEIs) expect students to critically read the Core Interactive Text (CIT) and review the provided media resources.

Applying DNA

- Core Interactive Text: Applying DNA
- Image: The Human Genome
- Video: Advantages of Genetically Modified Foods
- Video: Polymerase Chain Reaction
- Video: Growing Heartier Plants
- Video: Restriction Enzymes
- Image: PCR Components
- Video: Developments in Reproduction

TEACHER NOTE This is a formative assessment item. As this short writing assignment lets students discuss the ethics of bringing back extinct organisms, there is room for interpretation and opinion. Allow students to express their views, perhaps in the form of a class debate.

TEI Technology and Dinosaurs

STEM Project Starters

STEM Project Starters provide additional real-world contexts that require students to apply and extend their content knowledge related to the concept. STEM Project Starters can also serve as an alternative instructional hook presented at the beginning of the learning progression. The project can then be revisited throughout and at the end of the 5E learning cycle, for students to apply content knowledge.

STEM Project Starter: The "Other" Race for the Double Helix `Recommended 45 minutes`

What other model structures of DNA exist?

> **TEACHER NOTE** This formative STEM project relates the discovery of the DNA structure to the scientists involved with it. Linus Pauling, while most notable as a scientist for other discoveries, is a lesser-known player in the DNA arena.

TEI DNA Model

STEM Project Starter What Does DNA Look Like under a Microscope? `Recommended 45 minutes`

What can a microscope tell you about DNA?

> **TEACHER NOTE** Students should submit their reports at least twice—once when they have designed their procedure and once when they have conducted the investigation.

TEI DNA under a Microscope

STEM Project Starter How Far Does It Go? `Recommended 45 minutes`

What can gel electrophoresis tell you about DNA?

> **TEACHER NOTE** This is a formative project. It ties in the technology of gel electrophoresis with the structure of DNA. An extension of this project would be to have students actually run a sample if materials and equipment are available. They could then compare their results with those presented here.

TEI How Far Does It Go?

Evaluate 45–90 minutes

Explain Question:
How does DNA carry the code for a specific protein?

Lesson Questions (LQ):
- What are the components of DNA and RNA?
- How are the structures of DNA and RNA related to their functions?
- How do cells make copies of DNA?

Throughout instruction and the 5E learning cycle, you will have collected formative assessment data to drive the assignment of resources and experiences to students. Evaluate is intended to include summative assessment checks for proficiency. You can use the Explain and Lesson Questions for the concept as a summative assessment in a variety of ways such as these:

- Post each Lesson Question (LQ) in various locations in the classroom, and have small groups of students generate claim statements related to the Lesson Question (LQ). Other students can add to the claim, or refute the claim, during a gallery walk where they place additional pieces of evidence on each Lesson Question (LQ) poster.
- Assign small groups of students to each Lesson Question (LQ) and have the groups generate a poster, board, graphic, or piece of text that answers the question. Use a jigsaw approach and create a second set of groups that contain members from each Lesson Question (LQ) group to share their ideas.
- Ask students to return to their initial ideas for the Explain question and add additional details and evidence.

Encourage students to review the concept review and complete the Student Self-Check practice assessment prior to assigning the Summative Teacher Concept assessment.

- Student Review and Practice Assessment
- Teacher Concept Assessments

Transcription and Translation

The Five E Instructional Model

Science Techbook follows the 5E instructional model. This Model Lesson includes strategies for each of the 5Es. As you design the inquiry-based learning experience for students, be sure to collect data during instruction to drive your instructional decisions. Point-of-use teacher notes are also provided within each E-tab.

Engage   45–90 minutes

Engage Media Resources

The resources found in Engage are intended to stimulate students by exposing them to a phenomenon relevant to the content of the lesson. Engage also provides examples of relevant real-world applications that allow students to begin to make observations and relate the science content to their everyday lives. The Core Interactive Text (CIT) and media resources are carefully designed to prompt students to begin asking questions that they can investigate during the Explore phase of the lesson. They should also start collecting evidence to address the Explain question located at the bottom of the Engage page.

> **TEACHER NOTE** **Investigative Phenomenon:** The image of a model made from plastic bricks is used to get the students to think about the need to carefully read and carry out instructions when assembling a complex object. This analogy is then applied to the need for the cell to read and carefully copy the code carried by DNA and translate these into the instructions required to build a complex protein molecule.

- Core Interactive Text: Organism Instructions
- Image: Building an Organism
- Image: Complex Protein Structure
- Video: Sickle Cell Anemia
- Video: Cell City

Explain Question

The Explain question focuses students on gathering information in the Explore section. The Explain question can be used to

- Record what students already know related to the Explain question.
- Serve as a template or model for students to generate their own scientific questions.
- Collect evidence as students work through the lesson.
- Allow students to reflect on their growth before and after the lesson.

Explain

How is transcription related to translation by the steps in which different types of RNA are processed, thereby allowing cells to control gene expression, and how is gene expression affected by various types of mutations?

■ **Image** Variation in Eye Color

Engage Formative Assessment

Technology Enhanced Items (TEIs) found on the Engage page enable you to collect data on students' prior knowledge and identify the common misconceptions they may possess that are related to the topic of study. These items are designed as quick checks for understanding and allow each student one attempt at each question. You can use the data collected to decide whether to assign additional resources to the class, or determine what individual or groups of students may need reinforcement or accelerated learning, prior to completing the Explore portion of the lesson.

TEACHER NOTE Use this student response to evaluate students' prior knowledge of the concept. Students might confuse the process of transcription, in which the information in DNA is recoded into messenger RNA, with replication, in which copies of DNA with the same information are made. Be sure to address any misconception about translation, checking for understanding that proteins, not amino acids, are synthesized during translation.

TEI Genetics and Adoption

Before You Begin

What Do I Already Know about Transcription and Translation?

TEACHER NOTE This formative assessment is intended to provide the teacher with feedback on students' prior knowledge of this topic. In middle school, students should have learned about the differences in structure between the two nucleic acids involved with protein synthesis. Students may have misconceptions regarding the composition of DNA and mRNA and may not realize that mRNA forms a complementary strand to the DNA. Present this activity to the students as a class and then encourage them to work together to sort the items.

TEI DNA vs. RNA Structure

TEACHER NOTE This formative assessment reviews the process of DNA replication. It should be presented to students as a pretest before reviewing the topic. Students should answer the question individually. Students selecting A are probably incorrectly remembering DNA helicase. Students selecting C probably have the misconception that the bases are placed into alphabetical order before they are copied. This does not happen. Students selecting D may think that mutations are the result of the absence of DNA polymerase.

TEI DNA Replication

TEACHER NOTE This formative assessment provides the teacher with feedback on prior knowledge of this topic. Its purpose is to review the concepts of heredity. Students should know that the production of specific proteins results in the expression of particular traits. These traits are then passed on because the genes are replicated in parental gametes. The teacher should present this activity to the students as an entire class and then allow them to work in pairs to answer the questions.

TEI **Basics of Heredity**

TEACHER NOTE This formative assessment reviews misconceptions that students may have about how DNA and proteins are related and the role of translation. Some students may think that DNA produces proteins directly, forgetting about RNA. Provide this activity to students to complete individually and then discuss their answers as an entire class.

TEI **DNA**

- Video: The Role of Proteins
- Video: What Are DNA and RNA?
- Video: DNA Duplication
- Video: The Expression of Genes

Explore `180 minutes`

Lesson Questions (LQs):

1. What is transcription?
2. What are the steps involved in RNA processing?
3. What are the functions of messenger RNA (mRNA), transfer RNA (tRNA), and ribosomal RNA (rRNA)?
4. What is translation?
5. How do cells control gene expression?
6. What are the causes, types, and effects of mutations?

Effective science instruction involves a student-centered rather than a teacher-centered approach. This can be accomplished either with Directed Inquiry or Guided Inquiry, depending on the needs and abilities of your class. Encourage students to select a variety of resources in their pursuit of answers as they work through Explore, with the end goal of constructing their scientific explanation in the Explain tab.

Directed Inquiry	Guided Inquiry
In Directed Inquiry, teachers provide students with a sequence of specified resources, challenging questions, and clear outcomes. Within this context students are given the opportunity to interact independently with each resource as prescribed by the teacher. Often different students groups can be guided through several different resources at the same time. For example, one group could work on a reading passage while a second group conducts a small-group Hands-On Activity with the teacher, and a third group is independently engaged with an online interactive resource.	In Guided Inquiry, students have independence to decide the scope and sequence of their investigations. Using resources from Tech book, students determine for themselves which resources they will Explore to answer the Lesson Questions. It is important to note that each student will choose multiple resources, but no one student is expected to use all the resources available. Students also determine the order in which to explore these resources and how to record their findings.

NGSS Components

SEP	CCC
■ Analyzing and Interpreting Data ■ Constructing Explanations and Designing Solutions	■ Structure and Function

Lesson Question: What Is Transcription?

`Recommended 15 minutes`

TEACHER NOTE Connections: Crosscutting Concept: Structure and Function: In this concept, students investigate DNA and the properties and structures that make it possible for cells to complete the function of protein synthesis. Through activities and assets, they will investigate transcription and translation, examining how components are shaped and used, including how ribosomes translate the genetic code. They will analyze structures and functions of mRNA, tRNA, and rRNA, and explore how these relate to genetic expression and factors that can cause helpful or harmful genetic mutations. Use a strategy such as Jigsaw to help students understand how the different components and processes of protein synthesis work together. Organize groups, assign each member a different topic, regroup all students with the same topic, provide appropriate resources and time for students to research and ask questions, and finally have students return to their original groups to take turns sharing what they learned. For information on this strategy, go to the Professional Learning tab of Tech book, then click on Strategies & Resources, then click on Spotlight on Strategies.

- Core Interactive Text: What Is Transcription?
- Image: DNA Double Helix
- Video: Transcription

TEACHER NOTE Misconception: Students may confuse transcription and replication, thinking that mRNA is a copy of the DNA sequence. In fact, mRNA is a different type of nucleic acid. In RNA, uracil pairs with adenine whereas in DNA, thymine pairs with adenine. The code on the mRNA is the opposite of the DNA code from which it was assembled. However, the DNA code does indirectly code for the amino acid sequence of a protein. DNA codes for RNA that in turn codes for proteins.

Formative Assessment

Throughout Explore, Technology Enhanced Items (TEIs) are embedded as multi-dimensional formative checks for understanding. You can use the data they provide to

- assign additional support
- extend learning
- design additional learning tasks to clarify student misconceptions

The Explore TEIs provide students with three attempts to demonstrate their proficiency. Scaffolded feedback is provided for each attempt. If a student does not achieve proficiency by the third attempt, a media asset is provided as an additional learning opportunity.

TEACHER NOTE **Practices: Science and Engineering Practice: Analyzing and Interpreting Data:** In the following item, students will analyze data sets to determine first if they are consistent with the events that occur on a very small scale during transcription. Then they will organize the selected steps into the correct order and determine if the new data provides a working explanation of the process of transcription. This activity will assist them in differentiating this pattern from the pattern of translation. Extend this item by having students create a Venn diagram comparing and contrasting transcription and translation. Students often confuse these two processes and having a graphic organizer will help reinforce the similarities and differences.

TEI **Transcription**

Lesson Question: What Are the Steps Involved in RNA Processing?

`Recommended 10 minutes`

- Core Interactive Text: What Are the Steps Involved in RNA Processing?
- Image: Synthesis of mRNA
- Video: RNA and Transcription

Lesson Question: What Are the Functions of Messenger RNA (mRNA), Transfer RNA (tRNA), and Ribosomal RNA (rRNA)?

`Recommended 15 minutes`

- Core Interactive Text: What Are the Functions of Messenger RNA (mRNA), Transfer RNA (tRNA), and Ribosomal RNA (rRNA)?
- Video: Types of RNA

TEACHER NOTE **Connections: Crosscutting Concept: Structure and Function:** In the following item, the concept that the shape and stability of RNA and DNA are related to their functions will be reinforced as students describe the structures and functions of DNA and RNA to show similarities and differences. Then, they describe the functions of mRNA, tRNA, and rRNA to further demonstrate understanding of how more specialized structures function to connect the different steps in protein synthesis. Student answers will reveal a sense of scale as they describe the overall structure of the DNA double helix, the RNA single strand, and then the amino acids that make up these larger structures. To prepare for this item, have students create a tri fold comparing and contrasting the three molecules by completing the following:

- Draw a picture of each molecule
- Describe where it is found (nucleus/ribosome/both)
- Describe each function
- Write out the long versions of these abbreviations: mRNA, rRNA, tRNA
- Describe the processes each is involved in (transcription/translation/)

TEI **DNA and RNA**

Lesson Question: What Is Translation?   `Recommended 30 minutes`

- Core Interactive Text: What Is Translation?
- Video: The Central Dogma: Translation and the Code
- Image: mRNA Codons
- Video: Translation
- Image: Rough Endoplasmic Reticulum
- Video: Ribosomes and Protein Synthesis
- Exploration: Translation

TEACHER NOTE Practices: Science and Engineering Practice: Analyzing and Interpreting Data: In the following item, students analyze and interpret data using a genetic code chart as a model to determine the amino acid represented by each codon. Extend this item by having students come up with their own three-letter codons and finding the matching codon in the chart.

TEI Translate the Genetic Code

Lesson Question: How Do Cells Control Gene Expression?   `Recommended 65 minutes`

- Core Interactive Text: How Do Cells Control Gene Expression?
- Video: What Is Gene Expression?
- Video: Overview of Gene Expression
- Video: Human Chromosomes
- Reading Passage: The Five Discoveries of RNA Polymerase
- Video: Where Is Gene Expression Controlled?
- Image: Nucleic Acid Structure
- Exploration: Modeling DNA
- Video: Amphibian Development and Diversity
- Hands-On Lab: Transcription and Translation

Lesson Question: What Are the Causes, Types, and Effects of Mutations?   `Recommended 45 minutes`

- Core Interactive Text: What Are the Causes, Types, and Effects of Mutations?

TEACHER NOTE Students are generally interested in mutation. Extend thinking by having students complete either the Hands-On Activity: Investigating DNA or the Hands-On Activity: Researching Genetic Traits in Explore More Resources. Alternatively, students can read the passage Mix, Match, and Evolve! in Explore More Resources and discuss positive mutation in plant and animal breeding and related genetic engineering. Students can compare and contrast random gene mutation with genetic engineering or discuss current works of popular fiction that feature genetic mutation and discuss whether they could have happened, as well as how and if they would be adaptive.

- Video Mutations
- Hands-On Lab: Changes in DNA
- Video Damage to DNA Leads to Mutation
- Video What Influences Gene Expression?
- Video Other Environmental Influences on Gene Expression

TEACHER NOTE Practices: Science and Engineering Practice: Constructing Explanations and Designing Solutions: In the following item, students select correct terms to construct an explanation of the functions and structures of different components of gene expression and of how malfunctions in the structures or functions of translation can be harmful or beneficial to an organism. Create an illustration of the step-by-step process of protein synthesis with captions. Ask questions such as: How would you describe the sequence of gene expression? What can go wrong in the process? After the item, have students create and exchange a mutation problem using these steps:

- Student creates a DNA sequence with 15 bases.
- Student creates a mutated DNA sequence (insertion, deletion, substitution).
- Partner will determine mRNA sequence for both normal and mutated DNA.
- Partner will determine amino acid sequence for normal and mutated DNA.
- Partner will determine the type of mutation

TEI All About Gene Expression

Explore More Resources

Resources in Explore More Resources support differentiation within your classroom by

- providing additional visualization of content
- affording extension of content to those students ready for acceleration
- offering Lexile reading levels for reading passages

Online explorations and hands-on experiences are provided so that students can conduct virtual investigations, collect and design investigations, and collect and analyze data; these skills are essential to developing scientific understanding.

Explain   45–90 minutes

In Explore, students
1. uncovered scientific understandings
2. conducted investigations
3. analyzed data, text, and other media resources
4. collected evidence to support their scientific explanation

In Explain, provide students with time to formally compose their scientific explanations around the Explain or student-generated questions using evidence collected from Explore.

Scientific explanations are student responses, either written or orally presented, that explain scientific phenomena based upon evidence. Developing a scientific explanation requires students to analyze and interpret data to construct meaning out of the data. There are three main components to the scientific explanation: the claim, the evidence, and the reasoning.

To help students to communicate their scientific explanations, allow them to utilize the multimedia creation tools such as Board Builder and Whiteboard. Remind them that they may upload image, audio, and video files using the "attach file" option to communicate their scientific explanations.

Students may construct their scientific explanations individually or within a small group of students. Students should communicate their explanations with other classmates, and provide constructive criticism and refine their explanations prior to submission to the teacher. If explanations are used as a formative assessment, you can provide additional feedback and comments to support students as they refine their explanations.

CAN YOU EXPLAIN?

How is transcription related to translation by the steps in which different types of RNA are processed, thereby allowing cells to control gene expression, and how is gene expression affected by various types of mutations?

Elaborate with STEM `45–135 minutes`

*Elaborate with STEM are optional extension resources available after students have demonstrated proficiency with standards addressed previously in the concept.

NGSS Components

SEP	CCC
■ Developing and Using Models ■ Analyzing and Interpreting Data ■ Using Mathematics and Computational Thinking ■ Constructing Explanations and Designing Solutions ■ Engaging in Argument from Evidence ■ Obtaining, Evaluating, and Communicating Information	■ Patterns ■ Cause and Effect ■ Structure and Function

STEM In Action `45 minutes`

STEM in Action ties the scientific concepts to real-world applications, with many connecting to STEM careers. Technology Enhanced Items (TEIs) expect students to critically read the Core Interactive Text (CIT) and review the provided media resources.

Applying Transcription and Translation

- Core Interactive Text: Applying Transcription and Translation
- Video: Genetic Diseases
- Image: Glowing Cancer Cells
- Video: Preventing Skin Cancer
- Video: Transgenic Animals for Agriculture
- Reading Passage: Treating Diabetes
- Video: Plans for Plasmids
- Video: The Genetic Engineering of Wheat
- Video: What Causes Cancer?
- Video: Transgenic Animals for Medicine

TEACHER NOTE This formative activity introduces students to the concept of mutation rate. Present it to the students after discussing the different types of mutations. Students should work indepencently on this assignment.

TEI Frequency of Mutations

UNIT 4: Inheritance and Variation

STEM Project Starters

STEM Project Starters provide additional real-world contexts that require students to apply and extend their content knowledge related to the concept. STEM Project Starters can also serve as an alternative instructional hook presented at the beginning of the learning progression. The project can then be revisited throughout and at the end of the 5E learning cycle, for students to apply content knowledge.

STEM Project Starter: The Frequency of a Mutation `Recommended 90 minutes`

How do changes during replication and translation influence the expression of new traits?

> **TEACHER NOTE** This formative project should be completed after students have completed the Hands-On Lab. Present this activity to students after the lab and use it to reinforce the concepts of gene expression and mutations. Present the activity to students as a class but then let them work individually on the report.

- Hands-On Lab: Variation and Distribution of Expressed Traits

`TEI` **Report**

STEM Project Starter: Explaining My Mutation `Recommended 45 minutes`

How does RNA's single-stranded structure provide an advantage when synthesizing proteins?

> **TEACHER NOTE** This summative project demonstrates how mutations are related to the expression of a particular trait. Students model the formation of a gene and then introduce the mutations to determine if a change in the trait will occur. They then compare their model of DNA to that of a classmate pair or group and determine the percent similarity. This activity should be done by pairs of students or larger groups, depending upon materials available.
>
> This project should be used in conjunction with the one titled "How to Synthesize Proteins," as that one delves more deeply into how amino acids joins together to form polypeptides. Example technologies that are used to detect mutations could include single-strand conformational polymorphism (SSCP) and denaturing gradient gel electrophoresis (DGGE). Various other techniques are available and should be accepted as answers as long as the student provides sufficient support from their research.

- Chart of Codons

`TEI` **Explaining My Mutation**

STEM Project Starter: How to Synthesize Proteins

Recommended 45 minutes

What molecules are involved in protein synthesis?

TEACHER NOTE This summative project can be used to assess students' understanding of how DNA codes for particular proteins. Students can work as an entire class or in groups, depending upon the materials available.

TEI **Presentation**

Evaluate `45–90 minutes`

Explain Question

How is transcription related to translation by the steps in which different types of RNA are processed, thereby allowing cells to control gene expression, and how is gene expression affected by various types of mutations?

Lesson Questions (LQ):

- What is transcription?
- What are the steps involved in RNA processing?
- What are the functions of messenger RNA (mRNA), transfer RNA (tRNA), and ribosomal RNA (rRNA)?
- What is translation?
- How do cells control gene expression?
- What are the causes, types, and effects of mutations?

Throughout instruction and the 5E learning cycle, you will have collected formative assessment data to drive the assignment of resources and experiences to students. Evaluate is intended to include summative assessment checks for proficiency. You can use the Explain and Lesson Questions for the concept as a summative assessment in a variety of ways such as these:

- Post each Lesson Question (LQ) in various locations in the classroom, and have small groups of students generate claim statements related to the Lesson Question (LQ). Other students can add to the claim, or refute the claim, during a gallery walk where they place additional pieces of evidence on each Lesson Question (LQ) poster.
- Assign small groups of students to each Lesson Question (LQ) and have the groups generate a poster, board, graphic, or piece of text that answers the question. Use a jigsaw approach and create a second set of groups that contain members from each Lesson Question (LQ) group to share their ideas.
- Ask students to return to their initial ideas for the Explain question and add additional details and evidence.

Encourage students to review the concept review and complete the Student Self-Check practice assessment prior to assigning the Summative Teacher Concept assessment.

- Student Review and Practice Assessment
- Teacher Concept Assessments

Genetic Disorders and Technology

The Five E Instructional Model

Science Techbook follows the 5E instructional model. This Model Lesson includes strategies for each of the 5Es. As you design the inquiry-based learning experience for students, be sure to collect data during instruction to drive your instructional decisions. Point-of-use teacher notes are also provided within each E-tab.

Engage   45–90 minutes

Engage Media Resources

The resources found in Engage are intended to stimulate students by exposing them to a phenomenon relevant to the content of the lesson. Engage also provides examples of relevant real-world applications that allow students to begin to make observations and relate the science content to their everyday lives. The Core Interactive Text (CIT) and media resources are carefully designed to prompt students to begin asking questions that they can investigate during the Explore phase of the lesson. They should also start collecting evidence to address the Explain question located at the bottom of the Engage page.

> **TEACHER NOTE** **Investigative Phenomenon:** Engage focuses on genetic disorders in humans. Have students look at the image titled "Down Syndrome." It shows a family. The boy in the family has the genetic disorder Down syndrome. Bear in mind that some students may have Down syndrome or have siblings that have Down syndrome. Ask the students: In what ways are the children different from the parents? Which individual exhibits characteristics that stand out from the others? What characteristics are these? Explain to students that the boy has Down syndrome: a genetic disorder caused by an extra chromosome, 21. Ask them how they think an extra chromosome could be acquired. When could this happen in the reproductive process? Use their ideas as a segue into other mistakes that could occur in the reproductive process and that could lead to genetic disorders. Have pairs of students list their ideas and then have them share them with the class. Some suggested causes are the gain or loss of any of the chromosomes, errors in a nucleotide DNA sequence during replication, and errors that include copying chunks of chromosome more than once or missing out parts of a chromosome altogether.

- Core Interactive Text: Understanding Genetic Disorders
- Image: Down Syndrome
- Video: Genetic Disorders
- Video: Hunting Down Genes Involved in Disease

Explain Question

The Explain question focuses students on gathering information in the Explore section. The Explain question can be used to

- Record what students already know related to the Explain question.
- Serve as a template or model for students to generate their own scientific questions.
- Collect evidence as students work through the lesson.
- Allow students to reflect on their growth before and after the lesson.

Explain

Explain how a person's genes affect their health.

Engage Formative Assessment

Technology Enhanced Items (TEIs) found on the Engage page enable you to collect data on students' prior knowledge and identify the common misconceptions they may possess that are related to the topic of study. These items are designed as quick checks for understanding and allow each student one attempt at each question. You can use the data collected to decide whether to assign additional resources to the class, or determine what individual or groups of students may need reinforcement or accelerated learning, prior to completing the Explore portion of the lesson.

TEI Genes and Your Health

Before You Begin

What Do I Already Know about Genetic Disorders and Heredity?

> **TEACHER NOTE** Use this student response as a formative assessment to prompt discussion and clarify common student misconceptions before you proceed with this topic. The Model Lesson provides information on common student misconceptions.
>
> Students should modify A to say that "Genetic disorders cannot be spread by sexual contact" and modify choice C to indicate that "Researchers have not found all possible connections between diseases and genetics."

TEI Genetic Disorders

> **TEACHER NOTE** This assessment item is intended to provide the teacher with feedback on prior knowledge of this topic. Use this item as a formative assessment to promote a class discussion.

TEI Punnett Square

TEACHER NOTE This formative assessment can provide the teacher with feedback on students' prior knowledge of this topic. Students can complete this item to assess understanding of protein synthesis before going forward with the rest of the topic. Use this activity as a think-pair-share activity.

TEI Protein Synthesis

TEACHER NOTE This formative activity is intended to provide the teacher with feedback on prior knowledge of this topic. Have students complete this before going forward with the rest of the topic to assess students' understanding of genetics. Use this activity to facilitate class discussion.

TEI Molecular Basis of Heredity

- Video: What Are DNA and RNA?
- Video: Gene Expression
- Video: Gamete Formation

Explore `135 minutes`

Lesson Questions (LQs):

1. What are the causes and symptoms of various genetic disorders?
2. What are the definitions and some examples of autosomal dominant, autosomal recessive, and sex-linked disorders?
3. What are the uses, benefits, and risks of genetic testing and gene therapy?

Effective science instruction involves a student-centered rather than a teacher-centered approach. This can be accomplished either with Directed Inquiry or Guided Inquiry, depending on the needs and abilities of your class. Encourage students to select a variety of resources in their pursuit of answers as they work through Explore, with the end goal of constructing their scientific explanation in the Explain tab.

Directed Inquiry	Guided Inquiry
In Directed Inquiry, teachers provide students with a sequence of specified resources, challenging questions, and clear outcomes. Within this context students are given the opportunity to interact independently with each resource as prescribed by the teacher. Often different students groups can be guided through several different resources at the same time. For example, one group could work on a reading passage while a second group conducts a small-group Hands-On Activity with the teacher, and a third group is independently engaged with an online interactive resource.	In Guided Inquiry, students have independence to decide the scope and sequence of their investigations. Using resources from Techbook, students determine for themselves which resources they will Explore to answer the Lesson Questions. It is important to note that each student will choose multiple resources, but no one student is expected to use all the resources available. Students also determine the order in which to explore these resources and how to record their findings.

NGSS Components

SEP	CCC
■ Developing and Using Models ■ Analyzing and Interpreting Data ■ Obtaining, Evaluating, and Communicating Information	■ Structure and Function

Lesson Question: What Are the Causes and Symptoms of Various Genetic Disorders?

Recommended 40 minutes

TEACHER NOTE **Practices: Science and Engineering Practice: Obtaining, Evaluating, and Communicating Information:** By the end of this explore, students will have critically read scientific literature adapted for classroom use. They will be able to describe symptoms, causes, and treatments of genetic disorders by paraphrasing them in simpler but still accurate terms. Throughout the exploration, have students graphically communicate their findings by creating a community bulletin board describing each genetic disorder, possible treatments, and people who are most commonly affected.

TEACHER NOTE **Misconception:** Students may believe that genetic disorders are acquired by individuals during their lifetime and relate this to the "genetic mutations" encountered in science fiction stories involving superheroes. Genetic disorders are either passed down from the parents or occur as a result of errors in meiosis.

- Core Interactive Text: What Are the Causes and Symptoms of Various Genetic Disorders?
- Image: Mutation During Replication
- Image: Normal and Sickled Blood Cells
- Image: Huntington Gene Mutation

TEACHER NOTE **Connections: Crosscutting Concept: Structure and Function:** Have students work in pairs or small groups to draw the structures of the original and faulty proteins illustrated in the videos. Students should write a caption for each illustration that describes how the change in the structure of the protein will alter the function of the protein. As an extension to this activity, have students use small building toys to create a model of a functional and nonfunctional protein so they can visually understand the connection between structure and function. Understanding the complementary puzzle-like properties of proteins is critical to understanding the damage that a wrongly shaped protein can inflict.

- Video: Damage to DNA Leads to Mutation
- Video: Three Types of Genetic Diseases
- Video: Sickle Cell Anemia
- Image: Child with Down Syndrome
- Image: Trisomy 21
- Video: Aneuploidy
- Video: New Drug May Be Key to Preventing Alzheimer's Disease

Formative Assessment

Throughout Explore, Technology Enhanced Items (TEIs) are embedded as multi-dimensional formative checks for understanding. You can use the data they provide to

- assign additional support
- extend learning
- design additional learning tasks to clarify student misconceptions

The Explore TEIs provide students with three attempts to demonstrate their proficiency. Scaffolded feedback is provided for each attempt. If a student does not achieve proficiency by the third attempt, a media asset is provided as an additional learning opportunity.

TEACHER NOTE **Practices: Science and Engineering Practice: Developing and Using Models:** In this item, students will examine a model to identify various types of gene and chromosomal mutations. Ask students to predict the relationship between these DNA changes and gene therapy. Extend this item by having students work in groups to develop their own models, cutting chromosomes out of colored paper. Each group should physically manipulate their chromosomes to represent one of the abnormalities in the item. Have groups show each other their chromosome models.

TEI Types of Chromosomal Mutations

Lesson Question: What Are the Definitions and Some Examples of Autosomal Dominant, Autosomal Recessive, and Sex-Linked Disorders?

Recommended 40 minutes

- Core Interactive Text: What Are the Definitions and Some Examples of Autosomal Dominant, Autosomal Recessive, and Sex-Linked Disorders?
- Image: Autosomal Dominant Inheritance
- Video: The Genetics of a Pharaoh's Family
- Image: Neurofibroma on the Optic Nerve
- Image: Muscular Atrophy
- Video: Sex-Linked Traits
- Image: Sex-Linked Color Blindness

TEACHER NOTE Give students more experience with the human chromosomes, homologous chromosomes, and chromosomal abnormalities with the Hands-On Lab: Human Karyotypes, found in Even More Resources. Students will pair chromosome sets for normal male, normal female, Down Syndrome, and Turner Syndrome.

- Exploration: Genetic Disorders and Technology

Lesson Question: What Are the Uses, Benefits, and Risks of Genetic Testing and Gene Therapy?

Recommended 55 minutes

- Core Interactive Text: What Are the Uses, Benefits, and Risks of Genetic Testing and Gene Therapy?
- Image: Nurse Performs Amniocentesis Test
- Video: Genetic Testing
- Image: Gene Therapy Using Viruses
- Video: Gene Therapy Research
- Video: Problems with Gene Therapy
- Video: Using Viruses in Gene Therapy
- Video: The Human Genome Project
- Video: The Human Genome Project: The Future
- Video: The Human Genome Project: Accomplishments
- Video: How is a Genome Sequenced?

TEACHER NOTE Practices: Science and Engineering Practice: Analyzing and Interpreting Data: In these items, students are asked to analyze and interpret the data collected in two tables about Alpha-1 antitrypsin deficiency, including the limitations of data analysis, such as measurement error, human error, and sample selection. Extend this item by asking the students one final question: If the frequency of alleles is calculated by the Hardy Weinberg equation $(p + q)^2$ for two alleles, what would be the base calculation for the number of alleles shown in the table? [Answer = 6 including the other column (x) so approximately $(m1 + m2 + m3 + s + z + x)^2$] The Hardy Weinberg principle is reviewed in the section on Evolution.

TEI ATTD Allele Frequency
TEI ATTD Mutations
TEI ATTD Questionnaire

Explore More Resources

Resources in Explore More Resources support differentiation within your classroom by

- providing additional visualization of content
- affording extension of content to those students ready for acceleration
- offering Lexile reading levels for reading passages

Online explorations and hands-on experiences are provided so that students can conduct virtual investigations, collect and design investigations, and collect and analyze data; these skills are essential to developing scientific understanding.

Explain `45–90 minutes`

In Explore, students
1. uncovered scientific understandings
2. conducted investigations
3. analyzed data, text, and other media resources
4. collected evidence to support their scientific explanation

In Explain, provide students with time to formally compose their scientific explanations around the Explain or student-generated questions using evidence collected from Explore.

Scientific explanations are student responses, either written or orally presented, that explain scientific phenomena based upon evidence. Developing a scientific explanation requires students to analyze and interpret data to construct meaning out of the data. There are three main components to the scientific explanation: the claim, the evidence, and the reasoning.

To help students to communicate their scientific explanations, allow them to utilize the multimedia creation tools such as Board Builder and Whiteboard. Remind them that they may upload image, audio, and video files using the "attach file" option to communicate their scientific explanations.

Students may construct their scientific explanations individually or within a small group of students. Students should communicate their explanations with other classmates, and provide constructive criticism and refine their explanations prior to submission to the teacher. If explanations are used as a formative assessment, you can provide additional feedback and comments to support students as they refine their explanations.

EXPLAIN
Explain how a person's genes affect their health.

Elaborate with STEM `45–135 minutes`

*Elaborate with STEM are optional extension resources available after students have demonstrated proficiency with standards addressed previously in the concept.

NGSS Components

SEP	CCC
■ Constructing Explanations and Designing Solutions ■ Engaging in Argument from Evidence ■ Obtaining, Evaluating, and Communicating Information	■ Structure and Function

STEM In Action `45 minutes`

STEM in Action ties the scientific concepts to real-world applications, with many connecting to STEM careers. Technology Enhanced Items (TEIs) expect students to critically read the Core Interactive Text (CIT) and review the provided media resources.

Applying Genetic Disorders and Technology

- Core Interactive Text: Applying Genetic Disorders and Technology
- Video: What Is Gene Therapy?
- Video: Fighting Cancer with Gene Therapy
- Image: STEM and Genetic Disorders and Technology
- Video: Harnessing the Power of the Gene
- Video: Manipulating Genes to Affect Human Traits
- Video: Current GMO Research

TEACHER NOTE Students' responses indicate their understanding of what SNPs are, and how genes are inherited. Students can refer back to the information in the lesson about SNP if needed to develop their responses. Use as a formative assessment to gauge student understanding of SNPs and how genes are inherited.

- **TEI** SNPs and Disorders
- **TEI** Inheritance
- **TEI** Genetic Variation

STEM Project Starters

STEM Project Starters provide additional real-world contexts that require students to apply and extend their content knowledge related to the concept. STEM Project Starters can also serve as an alternative instructional hook presented at the beginning of the learning progression. The project can then be revisited throughout and at the end of the 5E learning cycle, for students to apply content knowledge.

STEM Project Starter: Gene Therapy Research: **Recommended 60 minutes**

Can you provide convincing evidence to win an argument?

> **TEACHER NOTE** Use this project as a formative assessment for students to demonstrate an understanding of technology and develop the ability to reason to a conclusion using evidence. Before students begin working on their presentations, guide them in how the presentation should be structured. A strong presentation includes both an introduction and conclusion and uses a cohesive style throughout.
>
> Inform students that their reports must be structured to include an introduction, supportive paragraphs, and a conclusion. Connections between concepts and ideas should be cohesive. Introductions should clearly present the student's opinion, and there should be at least one paragraph in the body for every argument or claim that is used. Finally, the conclusion should summarize and restate the student's position.
>
> Give students the opportunity to perform peer reviews on one another's writing. Students can use this feedback to improve their transitions, language, style, introductions, and conclusions.

> **TEACHER NOTE** After students have completed this activity individually, organize a class debate where students present the reasons for their stance. Divide the class into two groups, one in support of the grant and one opposed to the grant. Alternate between the two groups, giving each student an opportunity to share one reason for their conclusion.

TEI Gene Therapy Grant

STEM Project Starter Become a Biomedical Engineer   **Recommended 90 minutes**

What is the role of RNA interference in treating cancer?

> **TEACHER NOTE** Use this activity as a formative assessment to assess students' understanding of the science and technology involved and their ability to design a potential solution as they assume the role of a biomedical engineer.

TEI Design a Cancer Treatment

- Activity: Engineering Design Sheet

Evaluate `45–90 minutes`

Explain Question:
Explain how a person's genes affect their health.

Lesson Questions (LQ):
- What are the causes and symptoms of various genetic disorders?
- What are the definitions and some examples of autosomal dominant, autosomal recessive, and sex-linked disorders?
- What are the uses, benefits, and risks of genetic testing and gene therapy?

Throughout instruction and the 5E learning cycle, you will have collected formative assessment data to drive the assignment of resources and experiences to students. Evaluate is intended to include summative assessment checks for proficiency. You can use the Explain and Lesson Questions for the concept as a summative assessment in a variety of ways such as these:

- Post each Lesson Question (LQ) in various locations in the classroom, and have small groups of students generate claim statements related to the Lesson Question (LQ). Other students can add to the claim, or refute the claim, during a gallery walk where they place additional pieces of evidence on each Lesson Question (LQ) poster.
- Assign small groups of students to each Lesson Question (LQ) and have the groups generate a poster, board, graphic, or piece of text that answers the question. Use a jigsaw approach and create a second set of groups that contain members from each Lesson Question (LQ) group to share their ideas.
- Ask students to return to their initial ideas for the Explain question and add additional details and evidence.

Encourage students to review the concept review and complete the Student Self-Check practice assessment prior to assigning the Summative Teacher Concept assessment.

- Student Review and Practice Assessment
- Teacher Concept Assessments

The Chemistry of Life

The Five E Instructional Model

Science Techbook follows the 5E instructional model. This Model Lesson includes strategies for each of the 5Es. As you design the inquiry-based learning experience for students, be sure to collect data during instruction to drive your instructional decisions. Point-of-use teacher notes are also provided within each E-tab.

Engage   45–90 minutes

Engage Media Resources

The resources found in Engage are intended to stimulate students by exposing them to a phenomenon relevant to the content of the lesson. Engage also provides examples of relevant real-world applications that allow students to begin to make observations and relate the science content to their everyday lives. The Core Interactive Text (CIT) and media resources are carefully designed to prompt students to begin asking questions that they can investigate during the Explore phase of the lesson. They should also start collecting evidence to address the Explain question located at the bottom of the Engage page.

> **TEACHER NOTE** **Investigative Phenomenon:** This lesson is about basic biochemistry. The building block of biological compounds is the element carbon. The investigative phenomenon for this lesson is the search for life beyond Earth. Will any extraterrestrial life forms that may be out there be carbon-based, or are there other elements that have carbon-like properties that could perform a similar role? What is it about carbon's structure and chemical properties that makes it such a special element? Molecules that contain carbon do not always have a biological origin. Can we be sure, when we survey a planet, moon, or comet, that any carbon we find was once part of a living thing? How are scientists approaching this problem?

- Core Interactive Text: Assessing Carbon for Life
- Video: The Hunt for Carbon
- Image: Versatility of Carbon

Explain Question

The Explain question focuses students on gathering information in the Explore section. The Explain question can be used to

- Record what students already know related to the Explain question.
- Serve as a template or model for students to generate their own scientific questions.
- Collect evidence as students work through the lesson.
- Allow students to reflect on their growth before and after the lesson.

Explain
What are some ways in which the structure of a compound influences its function?

■ Image: Organic Construction

Engage Formative Assessment
Technology Enhanced Items (TEIs) found on the Engage page enable you to collect data on students' prior knowledge and identify the common misconceptions they may possess that are related to the topic of study. These items are designed as quick checks for understanding and allow each student one attempt at each question. You can use the data collected to decide whether to assign additional resources to the class, or determine what individual or groups of students may need reinforcement or accelerated learning, prior to completing the Explore portion of the lesson.

TEACHER NOTE Use this summative student response to evaluate students' understanding of the material in this section. Students can complete this activity in pairs, or the activity could be used as class discussion.

TEI Life on Mars?

Before You Begin
What Do I Already Know about the Chemistry of Life?

TEACHER NOTE This formative activity is intended to provide feedback on prior knowledge of this topic. Activity should be completed individually. Students who select B may think that any carbon-containing compound is an organic compound. Students who select C may think that the term applies to any compounds in living organisms, while those who do not select D may think that the term organic refers only to compounds in living organisms.

TEI Organic Compounds

TEACHER NOTE Students should know the four main classes of organic biomolecules. Use individual responses to this formative activity to provide feedback on prior knowledge of the role of these molecules on the cellular level.

TEI Organic Functions

TEACHER NOTE Students may have heard of enzymes and some may have an understanding of their role in chemical reactions. This formative assessment activity will provide some feedback on individual student understanding of how enzymes work. The Model Lesson discusses some misconceptions. Students' individual responses can form the basis for class discussion on enzyme function.

TEI How Do Enzymes Work?

- Video: Basics of Atomic Structure
- Video: What Is in a Bond?
- Video: Chemical Reactions

Explore `180 minutes`

Lesson Questions (LQs):
1. What are the differences between inorganic and organic molecules?
2. What are the characteristics of carbohydrates, lipids, proteins, and nucleic acids?
3. What are the characteristics of enzymes?
4. What is the role of adenosine triphosphate (ATP) in cells?

Effective science instruction involves a student-centered rather than a teacher-centered approach. This can be accomplished either with Directed Inquiry or Guided Inquiry, depending on the needs and abilities of your class. Encourage students to select a variety of resources in their pursuit of answers as they work through Explore, with the end goal of constructing their scientific explanation in the Explain tab.

Directed Inquiry	Guided Inquiry
In Directed Inquiry, teachers provide students with a sequence of specified resources, challenging questions, and clear outcomes. Within this context students are given the opportunity to interact independently with each resource as prescribed by the teacher. Often different students groups can be guided through several different resources at the same time. For example, one group could work on a reading passage while a second group conducts a small-group Hands-On Activity with the teacher, and a third group is independently engaged with an online interactive resource.	In Guided Inquiry, students have independence to decide the scope and sequence of their investigations. Using resources from Techbook, students determine for themselves which resources they will Explore to answer the Lesson Questions. It is important to note that each student will choose multiple resources, but no one student is expected to use all the resources available. Students also determine the order in which to explore these resources and how to record their findings.

NGSS Components

SEP	CCC
■ Asking Questions and Defining Problems ■ Obtaining, Evaluating, and Communicating Information	■ Cause and Effect ■ Structure and Function

Lesson Question: What Are the Differences Between Inorganic and Organic Molecules?

Recommended 45 minutes

TEACHER NOTE **Practices: Science and Engineering Practice: Obtaining, Evaluating, and Communicating Information:** As students explore the chemistry of life, they will communicate scientific and technical information and ideas in multiple formats such as orally and textually. As the instructional hook for this concept, use the strategy "Can You Guess My 2-1-4." Using this strategy in your own classroom hooks students at the beginning of a class period by helping them connect important pieces of this concept together. To use this strategy, you will need to prepare two facts, one clue, and four pictures that will help introduce the concept to your students. Two suggested facts are 1. This concept discusses atoms, but it's not exactly physics. 2. This concept discusses something happening in your body right now. A good clue might be *This concept discusses what plants and animals have in common.* Some suggested images include Periodic Table; Solid, Liquid, Gas; and Deer in the Forest, all of which can be found in Even More Resources. After students have been presented with the information, have an open discussion about possible answers. This strategy is an effective way to have your students make inferences from the information they've been presented with and come to conclusions about new concepts or topics they will be learning. Access this strategy by clicking the Professional Learning tab on Science Techbook. Click Strategies & Resources, then Spotlight on Strategies (SOS). "Can You Guess My 2-1-4" is found underneath "Instructional Hooks."

- Core Interactive Text: What Are the Differences Between Inorganic and Organic Molecules?
- Image: Carbon Jewelry
- Video: Organic Compounds

TEACHER NOTE **Misconception:** Students may think that organic means "from living things" and inorganic means "not from living things." Many students relate the term organic with the organic produce found in grocery stores and think that organic has something to do with not using pesticides. In the context of science, *organic* refers to compounds that are carbon-based and that are composed primarily of carbon atoms bound to other types of atoms.

TEACHER NOTE **Misconception:** Students may think that all organic molecules are biological in origin or found in living organisms. However, many simple carbon compounds form in nature independently of biological systems, even in outer space.

Formative Assessment:

Throughout Explore, Technology Enhanced Items (TEIs) are embedded as multi-dimensional formative checks for understanding. You can use the data they provide to

- assign additional support
- extend learning
- design additional learning tasks to clarify student misconceptions

The Explore TEIs provide students with three attempts to demonstrate their proficiency. Scaffolded feedback is provided for each attempt. If a student does not achieve proficiency by the third attempt, a media asset is provided as an additional learning opportunity.

> **TEACHER NOTE** Practices: Science and Engineering Practice: Asking Questions and Defining Problems: To understand and apply knowledge, students must first ask questions that arise from careful observations of phenomena, such as those that determine relationships like the difference between organic and inorganic molecules. Have students examine the question, "How can organic and inorganic molecules be classified?" Have them make a flow chart of questions that will help differentiate molecules, and then add to it as they move through this concept. (Questions on the flow chart may include "What is the molecule's formula?" "Does it contain carbon?" and "Is it a mineral?") After examining the molecules in this item, ask students what additional information they need to seek or clarify to better differentiate molecules.

TEI Organic or Inorganic?

Lesson Question: What Are the Characteristics of Carbohydrates, Lipids, Proteins, and Nucleic Acids?

`Recommended 45 minutes`

- Core Interactive Text: What Are the Characteristics of Carbohydrates, Lipids, Proteins, and Nucleic Acids?
- Image: Common Carbohydrates
- Hands-On Activity: Forming Carbohydrates
- Video: Carbohydrates
- Video: Lipids
- Image: Adipose Tissue
- Image: Molecular Structure of Fats
- Video: Fatty Acids
- Video: Proteins
- Video: Protein Synthesis
- Image: DNA Structure
- Video: Chemical Structure of DNA
- Video: Nucleic Acids
- Video: The DNA Code
- Image: The 46 Human Chromosomes
- Video: Carbohydrates

Lesson Question: What Are the Characteristics of Enzymes?

`Recommended 45 minutes`

- Core Interactive Text: What Are the Characteristics of Enzymes?
- Video: What Are Enzymes?

UNIT 5: Cells to Organisms

TEACHER NOTE Connections: Crosscutting Concept: Structure and Function: As students read the next section of CIT text, use the SOS strategy Scrambled Please as a way to sequence the steps. Access this strategy by clicking the Professional Learning tab on Science Techbook. Click Strategies & Resources, then Spotlight on Strategies (SOS). Scrambled Please is found underneath Sequence.

As students investigate systems, the properties and structures of the different components, and their interconnections, they will infer the functions and properties of natural and designed objects and systems from their overall structure, the way their components are shaped and used, and the molecular substructures of their various materials. This is especially true for the "lock and key model" of enzyme action.

TEACHER NOTE Misconception: Students may think that enzymes catalyze or trigger chemical reactions that would not normally take place on their own. In fact, enzymes catalyze reactions that will proceed on their own and serve to improve the efficiency of these reactions.

- Video: Enzymes
- Video: How Enzymes Catalyze Reactions
- Video: Effect of pH on Enzymes
- Video: Enzymes and Coenzymes

TEACHER NOTE Connections: Crosscutting Concept: Cause and Effect: In this item, students make claims about specific causes and effects to explain behaviors in complex natural systems, including genes and enzymes, and complex designed systems, such as the focus of genetic researchers. The causal relationships at the level of the individual organism rely upon an examination of what is known about smaller scale mechanisms within the system, at the level of enzymes, proteins, and genes. To extend the item, divide the students into five groups and use the strategy 25 Things You Didn't Know to explore what is known about the causes of genetic diseases; knowledge about genes, DNA, and proteins; and the effect of this knowledge in terms of treatments and future cures. Good resource videos include Howard Hughes Medical Institute: Sickle Cell Anemia and Three Types of Genetic Diseases, found in Even More Resources. When groups finish creating their list, have each group share their five facts and discuss why they chose those facts. Challenge students to differentiate between correlation and causation in the course of this discussion. Access 25 Things You Didn't Know by clicking the Professional Learning tab on Science Techbook. Click Strategies & Resources, then Spotlight on Strategies (SOS), and the Research section.

TEI Genetic Mapping

Lesson Question What Is the Role of Adenosine Triphosphate (ATP) in Cells?

Recommended 45 minutes

- Core Interactive Text: What Is the Role of Adenosine Triphosphate (ATP) in Cells?
- Image: Energy from ATP
- Video: ATP and Chemical Energy
- Video: Active Transport Processes

TEACHER NOTE Practices: Science and Engineering Practice: Obtaining, Evaluating, and Communicating Information: Before students attempt these items, make sure they have viewed the video segment Active Transport Processes. As the students complete the selected response items, they will compare a video segment, the text, and two images and then evaluate the information to address the scientific question of the cause of a rare disorder known as rapid-onset cystonia-parkinsonism (RDP). Dystonia is a condition characterized by spasmodic muscle movements. The students should be able to integrate all their previous knowledge with what they have learned in this chapter, and leverage it toward solving a real-life example.

To extend these items, use the SOS strategy Puppet Pictures. Have the students cut out images from the sodium potassium pump to include sodium, potassium, ATP, the cell membrane, and transport proteins and act out what happens both in a normal situation and in RDP. Access this strategy by clicking the Professional Learning tab on Science Techbook. Click Strategies & Resources, then Spotlight on Strategies (SOS). Puppet Pictures is found underneath Vocabulary Development.

TEI Cause of Disorder

TEI Onset of Disorder

Explore More Resources

Resources in Explore More Resources support differentiation within your classroom by

- providing additional visualization of content
- affording extension of content to those students ready for acceleration
- offering Lexile reading levels for reading passages

Online explorations and hands-on experiences are provided so that students can conduct virtual investigations, collect and design investigations, and collect and analyze data; these skills are essential to developing scientific understanding.

Explain `45–90 minutes`

In Explore, students
1. uncovered scientific understandings
2. conducted investigations
3. analyzed data, text, and other media resources
4. collected evidence to support their scientific explanation

In Explain, provide students with time to formally compose their scientific explanations around the Explain or student-generated questions using evidence collected from Explore.

Scientific explanations are student responses, either written or orally presented, that explain scientific phenomena based upon evidence. Developing a scientific explanation requires students to analyze and interpret data to construct meaning out of the data. There are three main components to the scientific explanation: the claim, the evidence, and the reasoning.

To help students to communicate their scientific explanations, allow them to utilize the multimedia creation tools such as Board Builder and Whiteboard. Remind them that they may upload image, audio, and video files using the "attach file" option to communicate their scientific explanations.

Students may construct their scientific explanations individually or within a small group of students. Students should communicate their explanations with other classmates, and provide constructive criticism and refine their explanations prior to submission to the teacher. If explanations are used as a formative assessment, you can provide additional feedback and comments to support students as they refine their explanations.

EXPLAIN

What are some ways in which the structure of a compound influences its function?

Elaborate with STEM `45–135 minutes`

*Elaborate with STEM are optional extension resources available after students have demonstrated proficiency with standards addressed previously in the concept.

NGSS Components

SEP	CCC
■ Asking Questions and Defining Problems ■ Planning and Carrying Out Investigations ■ Analyzing and Interpreting Data ■ Using Mathematics and Computational Thinking ■ Constructing Explanations and Designing Solutions Obtaining, Evaluating, and Communicating Information	■ Cause and Effect ■ Scale, Proportion, and Quantity ■ Structure and Function

STEM In Action `45 minutes`

STEM in Action ties the scientific concepts to real-world applications, with many connecting to STEM careers. Technology Enhanced Items (TEIs) expect students to critically read the Core Interactive Text (CIT) and review the provided media resources.

Applying Chemistry of Life

- Core Interactive Text: Applying Chemistry of Life
- Video: Enzymes in DNA Research and Industry
- Image: Glowing Fish
- Video: Proteins in 3-D
- Video: The Medication Detective

> **TEACHER NOTE** Use individual student responses on this activity as a summative assessment of student understanding of the material in this lesson.

TEI Designing Drugs

STEM Project Starters

STEM Project Starters provide additional real-world contexts that require students to apply and extend their content knowledge related to the concept. STEM Project Starters can also serve as an alternative instructional hook presented at the beginning of the learning progression. The project can then be revisited throughout and at the end of the 5E learning cycle, for students to apply content knowledge.

STEM Project Starter For Lack of an Enzyme   `Recommended 90 minutes`

How can biotechnology potentially address enzyme deficiency?

> **TEACHER NOTE** This formative activity could be completed individually or in pairs. The purpose of this project is to assess student understanding of the role of enzymes. Remind students that biochemical pathways have multiple steps and branches. Students should understand that there are two potential problems with enzyme deficiency: a buildup of substrate, and a lack of product.
>
> Inform students that their reports must be structured to include an introduction, supportive paragraphs, and a conclusion. Connections between concepts and ideas should be cohesive. Reports should include three supportive paragraphs that explain the function of the enzyme involved, how the enzyme deficiency affects a person's health, and how the condition can be treated.
>
> Give students the opportunity to perform peer reviews on one another's writing. Students can use this feedback to improve their transitions, language, style, introductions, and conclusions.

TEI Enzyme Deficiencies

- Activity: Engineering Design Sheet

STEM Project Starter: What Stops an Enzyme?   `Recommended 90 minutes`

Do potatoes contain an enzyme that neutralizes hydrogen peroxide?

> **TEACHER NOTE** This formative STEM project should be completed in pairs. Use the students' submissions as the basis for a class discussion on the structure and function. If students have difficulty identifying another enzyme, one possibility is amylase in saliva and its effect on starch.

- Hands-On Lab: Effect of Temperature and pH on Enzymes

TEI What Stops an Enzyme?

STEM Project Starter: Beverage Detective   `Recommended 60 minutes`

Do you know what's in your favorite beverage?

> **TEACHER NOTE** Use this formative project to assess student understanding of nutrition labels and chemical components of beverages.
>
> Students can complete the activity in pairs. As an extension, compile class results and compare different types of beverages. This activity can lead to a discussion of the types of nutrients that make up a nutritious diet.

TEI A Standard Curve

TEI Calculations

5.1 The Chemistry of Life

Evaluate 45–90 minutes

Explain? Question:
What are some ways in which the structure of a compound influences its function?

Lesson Questions (LQ):
- What are the differences between inorganic and organic molecules?
- What are the characteristics of carbohydrates, lipids, proteins, and nucleic acids?
- What are the characteristics of enzymes?
- What is the role of adenosine triphosphate (ATP) in cells?

Throughout instruction and the 5E learning cycle, you will have collected formative assessment data to drive the assignment of resources and experiences to students. Evaluate is intended to include summative assessment checks for proficiency. You can use the Explain and Lesson Questions for the concept as a summative assessment in a variety of ways such as these:

- Post each Lesson Question (LQ) in various locations in the classroom, and have small groups of students generate claim statements related to the Lesson Question (LQ). Other students can add to the claim, or refute the claim, during a gallery walk where they place additional pieces of evidence on each Lesson Question (LQ) poster.
- Assign small groups of students to each Lesson Question (LQ) and have the groups generate a poster, board, graphic, or piece of text that answers the question. Use a jigsaw approach and create a second set of groups that contain members from each Lesson Question (LQ) group to share their ideas.
- Ask students to return to their initial ideas for the Explain question and add additional details and evidence.

Encourage students to review the concept review and complete the Student Self-Check practice assessment prior to assigning the Summative Teacher Concept assessment.

- Student Review and Practice Assessment
- Teacher Concept Assessments

Cell Structure and Function

The Five E Instructional Model

Science Techbook follows the 5E instructional model. This Model Lesson includes strategies for each of the 5Es. As you design the inquiry-based learning experience for students, be sure to collect data during instruction to drive your instructional decisions. Point-of-use teacher notes are also provided within each E-tab.

Engage   45–90 minutes

Engage Media Resources

The resources found in Engage are intended to stimulate students by exposing them to a phenomenon relevant to the content of the lesson. Engage also provides examples of relevant real-world applications that allow students to begin to make observations and relate the science content to their everyday lives. The Core Interactive Text (CIT) and media resources are carefully designed to prompt students to begin asking questions that they can investigate during the Explore phase of the lesson. They should also start collecting evidence to address the Explain question located at the bottom of the Engage page.

> **TEACHER NOTE** **Investigative Phenomenon:** The investigative phenomenon for this concept links the behavior of grizzly bears to their biochemistry and cells. Grizzlies store fat in special cells that make up adipose tissue. How the fat helps the grizzly survive its winter hibernation relates to the fact that fat has a very high calorific value and insulating properties, both of which keep the bears warm during the winter. Use this phenomenon to help students make connections between structure and function at different scales in organisms. The focus of this concept is on sub-cellular structure. Later in this unit, students will look at how these cells, and see how their organelles, are organized to perform functions such as keeping organisms warm. Use the image of the adipose tissue (and other images of cells in Explore More Resources) to stimulate discussion and tap into students' existing knowledge on the appearance and size of cells and how they differ from each other.

- Core Interactive Text: Grizzly Fat Cells
- Image: Adipose Tissue under the Microscope
- Video: Cell Structure and Function

Explain Question

The Explain question focuses students on gathering information in the Explore section. The Explain question can be used to

- Record what students already know related to the Explain question.
- Serve as a template or model for students to generate their own scientific questions.
- Collect evidence as students work through the lesson.
- Allow students to reflect on their growth before and after the lesson.

Explain
What kind of structures must a cell have that would be analogous to a modern business?

- Image: An Automated Factory

Engage Formative Assessment
Technology Enhanced Items (TEIs) found on the Engage page enable you to collect data on students' prior knowledge and identify the common misconceptions they may possess that are related to the topic of study. These items are designed as quick checks for understanding and allow each student one attempt at each question. You can use the data collected to decide whether to assign additional resources to the class, or determine what individual or groups of students may need reinforcement or accelerated learning, prior to completing the Explore portion of the lesson.

TEACHER NOTE Use this student response to evaluate students' prior knowledge of the concept. The Model Lesson provides information on common student misconceptions.

TEI Structures a Cell Must Have

Before You Begin
What Do I Already Know about cell structure and function?

TEACHER NOTE This activity is intended to provide the teacher with feedback on prior knowledge of this topic. This should be used as a cooperative activity and be used for formative purposes.

TEI Prokaryotic and Eukaryotic

TEACHER NOTE This activity is intended to provide the teacher with feedback on prior knowledge of this topic. This activity should be used as an individual activity and for formative purposes.

TEI Plants and Animals

TEACHER NOTE This activity is intended to provide the teacher with feedback on prior knowledge of this topic. This activity should be used as a cooperative activity and be used for formative purposes.

TEI Cell Theory

- Video: Classification of Organisms
- Video: How to Prepare and Observe a Microscope Slide

Explore `180 minutes`

Lesson Questions (LQs):

1. What are the three parts of the cell theory?
2. What are the differences between prokaryotic and eukaryotic cells?
3. What structures make up cells?
4. What are the differences between plant and animal cells?
5. What is the role of mitochondria in eukaryotic cells?
6. What is the structure and function of the cell membrane?

Effective science instruction involves a student-centered rather than a teacher-centered approach. This can be accomplished either with Directed Inquiry or Guided Inquiry, depending on the needs and abilities of your class. Encourage students to select a variety of resources in their pursuit of answers as they work through Explore, with the end goal of constructing their scientific explanation in the Explain tab.

Directed Inquiry	Guided Inquiry
In Directed Inquiry, teachers provide students with a sequence of specified resources, challenging questions, and clear outcomes. Within this context students are given the opportunity to interact independently with each resource as prescribed by the teacher. Often different students groups can be guided through several different resources at the same time. For example, one group could work on a reading passage while a second group conducts a small-group Hands-On Activity with the teacher, and a third group is independently engaged with an online interactive resource.	In Guided Inquiry, students have independence to decide the scope and sequence of their investigations. Using resources from Techbook, students determine for themselves which resources they will Explore to answer the Lesson Questions. It is important to note that each student will choose multiple resources, but no one student is expected to use all the resources available. Students also determine the order in which to explore these resources and how to record their findings.

NGSS Components

SEP	CCC
■ Analyzing and Interpreting Data ■ Engaging in Argument from Evidence	■ Patterns ■ Structure and Function

Lesson Question What Are the Three Parts of the Cell Theory? **Recommended 25 minutes**

TEACHER NOTE **Connections: Crosscutting Concept: Structure and Function:** In this concept, students will explore the structure of cells and will understand how their parts function together. Students will examine the key parts and properties of cells and will question how they relate to each other. They will investigate cells by examining the structures of different organelles and their interconnections to reveal the overall function of the cell. Help students understand the structure and function of cells and cell parts by having them draw or write key concepts in their journal. The Journals strategy is found on the Professional Learning tab. Click on Strategies & Resources, then click on Spotlight On Strategies (SOS). Then click on Key Ideas and Details and Spotlight On Strategies: Journals.

- Core Interactive Text: What Are the Three Parts of the Cell Theory?
- Reading Passage: The Cell Theory
- Video: Cell Theory
- Image: Ostrich Egg
- Video: Cells: The Building Blocks of Life
- Image: Adipose Tissue

Formative Assessment:

Throughout Explore, Technology Enhanced Items (TEIs) are embedded as multi-dimensional formative checks for understanding. You can use the data they provide to

- assign additional support
- extend learning
- design additional learning tasks to clarify student misconceptions

The Explore TEIs provide students with three attempts to demonstrate their proficiency. Scaffolded feedback is provided for each attempt. If a student does not achieve proficiency by the third attempt, a media asset is provided as an additional learning opportunity.

TEACHER NOTE **Practices: Science and Engineering Practice: Engaging in Argument from Evidence:** In this item, students will construct an argument as to why humans meet the criteria of the cell theory using scientific evidence. To prepare for this activity, use the Hot Potato strategy to help students think about the components of the cell theory. Have them ask questions regarding what makes different living things meet the cell theory requirements. The Hot Potato strategy is found on the Professional Learning tab. Click on Strategies & Resources, then click on Spotlight On Strategies (SOS). Now click on Questioning, then click on Spotlight On Strategies: Hot Potato.

TEI **Cell Theory**

Lesson Question: What Are the Differences between Prokaryotic and Eukaryotic Cells?

Recommended 85 minutes

- Core Interactive Text: What Are the Differences between Prokaryotic and Eukaryotic Cells?
- Video: Introducing Cells
- Video: Prokaryotic and Eukaryotic Cells
- Hands-On Lab: Classifying Prokaryotic and Eukaryotic Cells

TEACHER NOTE Connections: Crosscutting Concept: Patterns: In this item, students distinguish between prokaryotic and eukaryotic cells. They observe patterns in systems at different scales and cite these patterns to distinguish between classifications. Have students use a graphic organizer to organize their understanding of prokaryotic and eukaryotic characteristics. Let them use words or pictures to help them remember the differences between these domains of life.

- Image: Prokaryotic and Eukaryotic

Lesson Question: What Structures Make Up Cells?

Recommended 25 minutes

- Core Interactive Text: What Structures Make Up Cells?
- Image: Bacteria
- Image: Bacterial Cell

TEACHER NOTE It may be difficult for students to keep track of the numerous subcellular structures found in cells. Encourage continued use of the Journals strategy by keeping a running list in the classroom of the name and function of each structure. Have students draw pictures to help them visualize the structures.

TEACHER NOTE Misconception: Students may be unaware of the complexity of cells or think that because they are very small, they have a very simple structure. This misconception is created in part by simple models or a simplistic diagrammatic representation of cells. Cells are very complex and dynamic. When using simplistic images to help explain cells, make sure you emphasize that they are simplified representations of real cells.

- Image: Animal Cell
- Video: The Organization of the Eukaryotic Cell
- Video: Components of the Nucleus
- Video: Endoplasmic Reticulum
- Video: Mitochondria, Centrosome, Endoplasmic Reticulum, and the Golgi Apparatus
- Video: Peroxisome, Lysosome, Chloroplast, and Vacuoles
- Image: Cytoskeleton
- Image: Trachea lining

Lesson Question: What Are the Differences between
Plant and Animal Cells?

Recommended 20 minutes

TEACHER NOTE Misconception: Students may think plant and animal cells are completely different from each other. In fact, while there are distinct differences between plant and animal cells, most structures found in plant cells are also found in animal cells.

- Core Interactive Text: What Are the Differences between Plant and Animal Cells?
- Exploration: Cell Structure and Function
- Video: Plant Cells Versus Animal Cells
- Video: Plant Cells
- Reading Passage: Onion and Cheek Cells

Lesson Question: What Is the Role of Mitochondria
in Eukaryotic Cells?

Recommended 10 minutes

- Core Interactive Text: What Is the Role of Mitochondria in Eukaryotic Cells?
- Image: Inside Mitochondria
- Video: Structure of Mitochondria
- Video: Mitochondria
- Video: Understanding ATP
- Video: Cellular Respiration

Lesson Question: What Is the Structure and Function
of the Cell Membrane?

Recommended 15 minutes

- Core Interactive Text: What Is the Structure and Function of the Cell Membrane?
- Video: Plasma Membrane
- Video: The Cell Membrane
- Image: Cell, Active Transport

TEACHER NOTE Practices: Science and Engineering Practice: Analyzing and Interpreting Data: In this item, students explain why a graph of active transport kinetics appear curved. They analyze data using a model in order to make valid and reliable scientific claims. This type of thinking may be challenging to students because it requires them to consider the implications of the active transport model. Support students by discussing the graph's axis labels and what an increase on each axis means. Extend this item and help students visualize active transport kinetics by building a life-sized model of a cell membrane using large cardboard boxes. Cut three doors to represent three channels in the cell membrane. Time and record data as, one, two, three, and more students enter the cell via the channels. Have students compare the data and evaluate why transport slows down as more and more students try to fit through the three doors. A smaller scale model could use a smaller box with doors sized for smaller objects, such as ping pong balls. The same experiment can be conducted.

TEI Transport Kinetics

Explore More Resources

Resources in Explore More Resources support differentiation within your classroom by

- providing additional visualization of content
- affording extension of content to those students ready for acceleration
- offering Lexile reading levels for reading passages

Online explorations and hands-on experiences are provided so that students can conduct virtual investigations, collect and design investigations, and collect and analyze data; these skills are essential to developing scientific understanding.

Explain `45–90 minutes`

In Explore, students
1. uncovered scientific understandings
2. conducted investigations
3. analyzed data, text, and other media resources
4. collected evidence to support their scientific explanation

In Explain, provide students with time to formally compose their scientific explanations around the Explain or student-generated questions using evidence collected from Explore.

Scientific explanations are student responses, either written or orally presented, that explain scientific phenomena based upon evidence. Developing a scientific explanation requires students to analyze and interpret data to construct meaning out of the data. There are three main components to the scientific explanation: the claim, the evidence, and the reasoning.

To help students to communicate their scientific explanations, allow them to utilize the multimedia creation tools such as Board Builder and Whiteboard. Remind them that they may upload image, audio, and video files using the "attach file" option to communicate their scientific explanations.

Students may construct their scientific explanations individually or within a small group of students. Students should communicate their explanations with other classmates, and provide constructive criticism and refine their explanations prior to submission to the teacher. If explanations are used as a formative assessment, you can provide additional feedback and comments to support students as they refine their explanations.

EXPLAIN

What kind of structures must a cell have that would be analogous to a modern business?

Elaborate with STEM   `45–135 minutes`

*Elaborate with STEM are optional extension resources available after students have demonstrated proficiency with standards addressed previously in the concept.

NGSS Components

SEP	CCC
■ Asking Questions and Defining Problems ■ Developing and Using Models ■ Planning and Carrying Out Investigations ■ Constructing Explanations and Designing Solutions ■ Obtaining, Evaluating, and Communicating Information	■ Scale, Proportion, and Quantity ■ Systems and System Models ■ Structure and Function

STEM In Action   `45 minutes`

STEM in Action ties the scientific concepts to real-world applications, with many connecting to STEM careers. Technology Enhanced Items (TEIs) expect students to critically read the Core Interactive Text (CIT) and review the provided media resources.

Applying Cell Structure and Function

- Core Interactive Text: Applying Cell Structure and Function
- Image: Bacteria
- Video: Beneficial Bacteria
- Video: Cool Jobs in Science

TEACHER NOTE Refer to the Examining Cells and Tissues Teacher Guide.

- Hands-On Lab: Examining Cells and Tissues
- Video: Antoni van Leeuwenhoek
- Video: Creating Microscope

STEM Project Starters

STEM Project Starters provide additional real-world contexts that require students to apply and extend their content knowledge related to the concept. STEM Project Starters can also serve as an alternative instructional hook presented at the beginning of the learning progression. The project can then be revisited throughout and at the end of the 5E learning cycle, for students to apply content knowledge.

STEM Project Starter Make Your Own Microscope `Recommended 45 minutes`

Microsccpes can be very large and expensive, but how can you construct one using a cell phone?

TEACHER NOTE Students should submit their procedure before beginning. If students are unable to see the Elodea, have them go back and check their steps to make sure they set up their microscope correctly. This project can be used as either a summative or formative assessment.

■ Hands-On Lab: Examining Cells and Tissues

TEI Cell Phone Microscope

■ Activity: Engineering Design Sheet

STEM Project Starter Create a Model of a Cell Membrane `Recommended 90 minutes`

How can you construct a 3-D model of a cell membrane?

TEACHER NOTE This project can be used as a formative or summative assessment. It is an extension of the lab in which students created a model of the atom. This should be used as a basis for class discussion of the cell membrane.

TEI Attach Model

■ Activity: Engineering Design Sheet

Evaluate `45–90 minutes`

Explain Question:
What kind of structures must a cell have that would be analogous to a modern business?

Lesson Questions (LQ):
- What are the three parts of the cell theory?
- What are the differences between prokaryotic and eukaryotic cells?
- What structures make up cells?
- What are the differences between plant and animal cells?
- What is the role of mitochondria in eukaryotic cells?
- What is the structure and function of the cell membrane?

Throughout instruction and the 5E learning cycle, you will have collected formative assessment data to drive the assignment of resources and experiences to students. Evaluate is intended to include summative assessment checks for proficiency. You can use the Explain and Lesson Questions for the concept as a summative assessment in a variety of ways such as these:

- Post each Lesson Question (LQ) in various locations in the classroom, and have small groups of students generate claim statements related to the Lesson Question (LQ). Other students can add to the claim, or refute the claim, during a gallery walk where they place additional pieces of evidence on each Lesson Question (LQ) poster.
- Assign small groups of students to each Lesson Question (LQ) and have the groups generate a poster, board, graphic, or piece of text that answers the question. Use a jigsaw approach and create a second set of groups that contain members from each Lesson Question (LQ) group to share their ideas.
- Ask students to return to their initial ideas for the Explain question and add additional details and evidence.

Encourage students to review the concept review and complete the Student Self-Check practice assessment prior to assigning the Summative Teacher Concept assessment.

- Student Review and Practice Assessment
- Teacher Concept Assessments

Cell Transport

The Five E Instructional Model
Science Techbook follows the 5E instructional model. This Model Lesson includes strategies for each of the 5Es. As you design the inquiry-based learning experience for students, be sure to collect data during instruction to drive your instructional decisions. Point-of-use teacher notes are also provided within each E-tab.

Engage   45–90 minutes

Engage Media Resources
The resources found in Engage are intended to stimulate students by exposing them to a phenomenon relevant to the content of the lesson. Engage also provides examples of relevant real-world applications that allow students to begin to make observations and relate the science content to their everyday lives. The Core Interactive Text (CIT) and media resources are carefully designed to prompt students to begin asking questions that they can investigate during the Explore phase of the lesson. They should also start collecting evidence to address the Explain question located at the bottom of the Engage page.

> **TEACHER NOTE** **Investigative Phenomena:** In Engage, students are introduced to the role of the cell membrane as the gatekeeper to the cell. The analogy of airport security and baggage handling is used to introduce what happens to substances that arrive and enter the cell. After completing Engage, have students share their opinions regarding how accurately the baggage handling process models the behavior and characteristics of the cell membrane.

- Core Interactive Text: Exploring Cell Transport
- Video: Airport Baggage
- Video: Diffusion
- Video: Cell Membrane

Explain Question
The Explain question focuses students on gathering information in the Explore section. The Explain question can be used to

- Record what students already know related to the Explain question.
- Serve as a template or model for students to generate their own scientific questions.
- Collect evidence as students work through the lesson.
- Allow students to reflect on their growth before and after the lesson.

Explain
How do cells regulate the passage of materials across their membranes in order to maintain homeostasis?

> ■ Animation: Examining the Cell Membrane

Engage Formative Assessment
Technology Enhanced Items (TEIs) found on the Engage page enable you to collect data on students' prior knowledge and identify the common misconceptions they may possess that are related to the topic of study. These items are designed as quick checks for understanding and allow each student one attempt at each question. You can use the data collected to decide whether to assign additional resources to the class, or determine what individual or groups of students may need reinforcement or accelerated learning, prior to completing the Explore portion of the lesson.

> **TEACHER NOTE** Use this student response to evaluate students' prior knowledge of the concept. The Model Lesson provides information on common student misconceptions.
>
> This activity is a formative assessment.

TEI Your Ideas

Before You Begin
What Do I Already Know about Cell Transport?

> **TEACHER NOTE** This formative assessment item is intended to provide the teacher with feedback on prior knowledge of this topic. In middle school, students should have learned about the structure of the cell membrane and how its components work together. A classroom discussion of this item will allow the teacher to determine student familiarity with the needed vocabulary terms.

TEI The Cell Membrane

> **TEACHER NOTE** This activity can be used as part of a think-pair-share. Have students begin by generating a list of items that dissolve when they get wet. Have them share this information with their partners and see if they can determine common characteristics among the items.
>
> In this formative assessment item, students who select A might be unclear on the composition of a cell membrane. Students who select C could be confusing cell transport with the other functions of the cell membrane. Students who select D will likely need reinforcement of both basic cell transport and membrane structure concepts.

TEI Why Don't You Melt?

TEACHER NOTE In this formative activity, students show their understanding of the cell membrane by selecting which statements are correct. Selecting more than one statement indicates that students know specific details about the membrane and its functioning. Students should complete this activity on an individual basis.

TEI Role of the Cell Membrane

TEACHER NOTE This formative activity is intended to provide the teacher with feedback on prior knowledge of this topic. Its main focus is to determine whether students are aware of how the cell membrane structure is directly related to the ability of materials to pass across the membrane. This activity can be used as part of a group discussion; start the discussion by asking the students what would happen if a cell were hydrophilic instead of hydrophobic.

TEI Cell Membrane and Transport

- Video: Inside Cells
- Video: The Cell Membrane
- Video: Cell Structure and the Human Body

Explore `90 minutes`

Lesson Questions (LQs):

1. What is the difference between passive and active cell transport?

Effective science instruction involves a student-centered rather than a teacher-centered approach. This can be accomplished either with Directed Inquiry or Guided Inquiry, depending on the needs and abilities of your class. Encourage students to select a variety of resources in their pursuit of answers as they work through Explore, with the end goal of constructing their scientific explanation in the Explain tab.

Directed Inquiry	Guided Inquiry
In Directed Inquiry, teachers provide students with a sequence of specified resources, challenging questions, and clear outcomes. Within this context students are given the opportunity to interact independently with each resource as prescribed by the teacher. Often different students groups can be guided through several different resources at the same time. For example, one group could work on a reading passage while a second group conducts a small-group Hands-On Activity with the teacher, and a third group is independently engaged with an online interactive resource.	In Guided Inquiry, students have independence to decide the scope and sequence of their investigations. Using resources from Techbook, students determine for themselves which resources they will Explore to answer the Lesson Questions. It is important to note that each student will choose multiple resources, but no one student is expected to use all the resources available. Students also determine the order in which to explore these resources and how to record their findings.

NGSS Components

SEP	CCC
■ Developing and Using Models	■ Structure and Function

Lesson Question: What Is the Difference between Passive and Active Cell Transport?

`Recommended 90 minutes`

TEACHER NOTE Connections: Crosscutting Concept: Structure and Function: In this concept, students will explore the structure of proteins and membranes responsible for active and passive transport. They infer the functions and properties of natural systems from their overall structure, the way their components are shaped and used, and the molecular substructures of their various materials. Help students distinguish between active and passive transport with the help of a journal. As students encounter differences between the two types of transport, have them write or draw the differences on the graphic organizer. Access the Journals strategy by clicking on Professional Learning, then Strategies & Resources, then Spotlight on Strategies (SOS). The Journals Strategy is located underneath Key Ideas and Details.

- Core Interactive Text: What Is the Difference between Passive and Active Cell Transport?
- Image: Movement by Osmosis
- Hands-On Lab: Passive and Active Transport Maintain Homeostasis

TEACHER NOTE Misconception: Students may think that once molecules reach equilibrium in diffusion, the molecules stop moving. In fact, random molecular movements and collisions still occur. However, there is no longer a concentration gradient.

TEACHER NOTE Passive transport can be modeled with a dialysis membrane. Explain to students that a dialysis membrane has small nonselective pores in the membrane. Place food coloring in the membrane and place the membrane in a beaker of water. The food coloring should passively diffuse out of the membrane. Repeat the experiment with a plastic bag without pores. The food coloring should stay inside the bag.

- Video: Types of Passive Transport
- Exploration: Cell Transport
- Video: Transport Mechanisms Other Than Passive or Active
- Video: Active Transport Processes

Formative Assessment:

Throughout Explore, Technology Enhanced Items (TEIs) are embedded as multi-dimensional formative checks for understanding. You can use the data they provide to

- assign additional support
- extend learning
- design additional learning tasks to clarify student misconceptions

The Explore TEIs provide students with three attempts to demonstrate their proficiency. Scaffolded feedback is provided for each attempt. If a student does not achieve proficiency by the third attempt, a media asset is provided as an additional learning opportunity.

TEACHER NOTE **Practices: Science and Engineering Practice: Developing and Using Models:** In this item, students will use models to predict the role of transport in different cells and to predict the thermodynamics of active transport. They develop and use multiple types of models to provide mechanistic accounts and predict phenomena. Help students understand the concepts of active and passive transport with the "3 Truths . . . 1 Lie" strategy. Before they start this collection, provide three truths and one lie regarding active and passive transport. See if students can identify the lie. The "3 Truths . . . 1 Lie" strategy is found on the Professional Learning tab. Click on Strategies & Resources, then click on Spotlight on Strategies (SOS). Click on Key Ideas and Details, then click on Spotlight on Strategies: 3 Truths . . . 1 Lie.

TEI Low and High Solute Concentrations

TEI Active or Passive

Explore More Resources

Resources in Explore More Resources support differentiation within your classroom by

- providing additional visualization of content
- affording extension of content to those students ready for acceleration
- offering Lexile reading levels for reading passages

Online explorations and hands-on experiences are provided so that students can conduct virtual investigations, collect and design investigations, and collect and analyze data; these skills are essential to developing scientific understanding.

Explain `45–90 minutes`

In Explore, students
1. uncovered scientific understandings
2. conducted investigations
3. analyzed data, text, and other media resources
4. collected evidence to support their scientific explanation

In Explain, provide students with time to formally compose their scientific explanations around the CYE or student-generated questions using evidence collected from Explore.

Scientific explanations are student responses, either written or orally presented, that explain scientific phenomena based upon evidence. Developing a scientific explanation requires students to analyze and interpret data to construct meaning out of the data. There are three main components to the scientific explanation: the claim, the evidence, and the reasoning.

To help students to communicate their scientific explanations, allow them to utilize the multimedia creation tools such as Board Builder and Whiteboard. Remind them that they may upload image, audio, and video files using the "attach file" option to communicate their scientific explanations.

Students may construct their scientific explanations individually or within a small group of students. Students should communicate their explanations with other classmates, and provide constructive criticism and refine their explanations prior to submission to the teacher. If explanations are used as a formative assessment, you can provide additional feedback and comments to support students as they refine their explanations.

CAN YOU EXPLAIN?

How do cells regulate the passage of materials across their membranes in order to maintain homeostasis?

Elaborate with STEM `45–135 minutes`

*Elaborate with STEM are optional extension resources available after students have demonstrated proficiency with standards addressed previously in the concept.

NGSS Components

SEP	CCC
■ Asking Questions and Defining Problems ■ Planning and Carrying Out Investigations ■ Using Mathematics and Computational Thinking ■ Constructing Explanations and Designing Solutions ■ Obtaining, Evaluating, and Communicating Information	■ Cause and Effect ■ Structure and Function

STEM In Action `45 minutes`

STEM in Action ties the scientific concepts to real-world applications, with many connecting to STEM careers. Technology Enhanced Items (TEIs) expect students to critically read the Core Interactive Text (CIT) and review the provided media resources.

Applying Cell Transport

- Core Interactive Text: Applying Cell Transport
- Video: Hormones
- Video: Cancer and Its Treatments
- Video: Treatments for Kidney Disease
- Video: The Role of Kidneys
- Reading Passage: A Day in the Life of a Dialysis Technician
- Video: Cancer Cells

TEACHER NOTE This formative activity can be completed by pairs or small groups of students.

TEI How Fast Does It Go?

STEM Project Starters

STEM Project Starters provide additional real-world contexts that require students to apply and extend their content knowledge related to the concept. STEM Project Starters can also serve as an alternative instructional hook presented at the beginning of the learning progression. The project can then be revisited throughout and at the end of the 5E learning cycle, for students to apply content knowledge.

STEM Project Starter: Application of Cell Membranes   `Recommended 90 minutes`

How can cell membranes be used to further drug treatment research?

- Image: The Cell's Outer Layer

TEACHER NOTE Students should submit their reports on this formative activity at least twice—once when they have a draft of their essay and once after they have finished revising it.

- **TEI** Objective
- **TEI** Description
- **TEI** Results
- **TEI** Conclusions

STEM Project Starter Drinking the Ocean   `Recommended 90 minutes`

What technologies exist to convert saltwater into drinkable water?

TEACHER NOTE This formative activity will incorporate engineering skills and integrate them with the science behind desalination.

Many of the responses students will give are dependent upon the information they collect and how they design their models. The generalized rubric here should be modified to fit each teacher's needs.

- Video: Uses for Salt Water

- **TEI** Objective
- **TEI** Materials
- **TEI** Safety
- **TEI** Procedure
- **TEI** Analysis

- Activity: Engineering Design Sheet

Evaluate `45–90 minutes`

Explain? Question:

How do cells regulate the passage of materials across their membranes in order to maintain homeostasis?

Lesson Questions (LQ):

- What is the difference between passive and active cell transport?

Throughout instruction and the 5E learning cycle, you will have collected formative assessment data to drive the assignment of resources and experiences to students. Evaluate is intended to include summative assessment checks for proficiency. You can use the CYE and Lesson Questions for the concept as a summative assessment in a variety of ways such as these:

- Post each Lesson Question (LQ) in various locations in the classroom, and have small groups of students generate claim statements related to the Lesson Question (LQ). Other students can add to the claim, or refute the claim, during a gallery walk where they place additional pieces of evidence on each Lesson Question (LQ) poster.
- Assign small groups of students to each Lesson Question (LQ) and have the groups generate a poster, board, graphic, or piece of text that answers the question. Use a jigsaw approach and create a second set of groups that contain members from each Lesson Question (LQ) group to share their ideas.
- Ask students to return to their initial ideas for the Explain question and add additional details and evidence.

Encourage students to review the concept review and complete the Student Self-Check practice assessment prior to assigning the Summative Teacher Concept assessment.

- Student Review and Practice Assessment
- Teacher Concept Assessments

Cell Division

The Five E Instructional Model

Science Techbook follows the 5E instructional model. This Model Lesson includes strategies for each of the 5Es. As you design the inquiry-based learning experience for students, be sure to collect data during instruction to drive your instructional decisions. Point-of-use teacher notes are also provided within each E-tab.

Engage 45–90 minutes

Engage Media Resources

The resources found in Engage are intended to stimulate students by exposing them to a phenomenon relevant to the content of the lesson. Engage also provides examples of relevant real-world applications that allow students to begin to make observations and relate the science content to their everyday lives. The Core Interactive Text (CIT) and media resources are carefully designed to prompt students to begin asking questions that they can investigate during the Explore phase of the lesson. They should also start collecting evidence to address the Explain question located at the bottom of the Engage page.

> **TEACHER NOTE** **Investigative Phenomenon:** Students should understand the basic principle of cell division--that one cell divides into two. Use the video Burn Victim to probe students' prior knowledge of the role of cell division in growth and development. For example, they should be able to infer that skin can be grown in the laboratory because skin cells undergo division. To further probe students' understanding, ask students why the researchers need to use collagen as a scaffold. They should understand that without scaffolding, cells will grow haphazardly. The collagen provides a structure around which cells can grow, restoring tissue. Ask students to discuss their ideas about cell division and how division results in different kinds of cells. Use this discussion as scaffolding to guide students toward the Explain Question, in which students distinguish between eukaryotic and prokaryotic cell division. Consider returning to this investigative phenomenon after students have completed this concept to evaluate their ability to apply knowledge of cell division to real-world problems such as treating burns or other ailments such as cancer.

- Core Interactive Text: Thinking About Cell Division
- Video: Burn Victim
- Image: A Healing Burn Injury

Explain Question

The Explain question focuses students on gathering information in the Explore section. The Explain question can be used to

- Record what students already know related to the Explain question.
- Serve as a template or model for students to generate their own scientific questions.
- Collect evidence as students work through the lesson.
- Allow students to reflect on their growth before and after the lesson.

Explain

Explain what steps are involved in the cell division of a prokaryote versus a eukaryote.

- Image: Cell Division

TEACHER NOTE Use this student response to evaluate students' prior knowledge of the concept. The Model Lesson provides information on common student misconceptions. Use this activity as a whole-class activity.

TEI Your Ideas

Engage Formative Assessment

Technology Enhanced Items (TEIs) found on the Engage page enable you to collect data on students' prior knowledge and identify the common misconceptions they may possess that are related to the topic of study. These items are designed as quick checks for understanding and allow each student one attempt at each question. You can use the data collected to decide whether to assign additional resources to the class, or determine what individual or groups of students may need reinforcement or accelerated learning, prior to completing the Explore portion of the lesson.

TEACHER NOTE This activity is intended to provide the teacher with feedback on prior knowledge and misconceptions of this concept. Students should know that cell division differs between prokaryotes and eukaryotes. Use this formative assessment in a cooperative activity.

Before You Begin

What Do I Already Know About Cell Division?

TEI Prokaryotic vs. Eukaryotic

TEACHER NOTE This activity is intended to provide the teacher with feedback on prior knowledge and misconceptions of this concept. Students should know that chromosome packaging differs between prokaryotes and eukaryotes. Use this formative assessment in a cooperative activity.

TEI Chromosome Differences

TEACHER NOTE This activity is intended to provide the teacher with feedback on prior knowledge of this topic. Its main focus is to determine whether students are familiar with the cell cycle and the stages involved. This activity should be used as a whole-class activity and be used for formative purposes.

TEI **Describing the Cell Cycle**

- Video: Prokaryotic and Eukaryotic Cells
- Video: Genetics and Heredity

Explore `180 minutes`

Lesson Questions (LQs):

1. What is the structure of chromosomes?
2. What are the processes involved in cellular division?
3. What are the differences between cell division among prokaryotes and eukaryotes?
4. What is the cell cycle?
5. What are the characteristics of cancer cells?

Effective science instruction involves a student-centered rather than a teacher-centered approach. This can be accomplished either with Directed Inquiry or Guided Inquiry, depending on the needs and abilities of your class. Encourage students to select a variety of resources in their pursuit of answers as they work through Explore, with the end goal of constructing their scientific explanation in the Explain tab.

Directed Inquiry	Guided Inquiry
In Directed Inquiry, teachers provide students with a sequence of specified resources, challenging questions, and clear outcomes. Within this context, students are given the opportunity to interact independently with each resource as prescribed by the teacher. Often, different student groups can be guided through several different resources at the same time. For example, one group could work on a reading passage while a second group conducts a small-group Hands-On Activity with the teacher, and a third group is independently engaged with an online interactive resource.	In Guided Inquiry, students have independence to decide the scope and sequence of their investigations. Using resources from Techbook, students determine for themselves which resources they will Explore to answer the Lesson Questions. It is important to note that each student will choose multiple resources, but no one student is expected to use all the resources available. Students also determine the order in which to explore these resources and how to record their findings.

NGSS Components

SEP	CCC
■ Asking Questions and Defining Problems ■ Developing and Using Models	■ Systems and System Models ■ Stability and Change

Lesson Question: What Is the Structure of Chromosomes?　　**Recommended 45 minutes**

TEACHER NOTE Connections: Crosscutting Concept: Systems and System Models: In this concept, students will use models to understand the process of cell division. They use models to simulate the flow of matter and interactions within systems at different scales. Help students understand the process of mitosis by using the Visual Walkabout strategy. Put up pictures of each stage of the cell cycle around the classroom and allow students to ask questions about each image. The "Visual Walkabout" strategy is found on the Professional Learning tab. Click on Strategies & Resources, then click on Spotlight on Strategies (SOS). Now click on Compare and Contrast, then click on Spotlight on Strategies: Visual Walkabout.

- Core Interactive Text: What Is the Structure of Chromosomes?
- Image: Chromosomes
- Video: DNA Packaging

Formative Assessment:

Throughout Explore, Technology Enhanced Items (TEIs) are embedded as multi-dimensional formative checks for understanding. You can use the data they provide to

- assign additional support
- extend learning
- design additional learning tasks to clarify student misconceptions

The Explore TEIs provide students with three attempts to demonstrate their proficiency. Scaffolded feedback is provided for each attempt. If a student does not achieve proficiency by the third attempt, a media asset is provided as an additional learning opportunity.

TEACHER NOTE Practices: Science and Engineering Practice: Asking Questions and Defining Problems: In this item, students analyze questions about the structure of chromosomes to determine which questions are testable. They examine models of the structure of chromosomes to clarify and seek additional information and relationships. Help students understand the structure of chromosomes by building a chromosome model with them. Use yarn for DNA and beads for histones and have them wind the DNA around beads and twist the chromatin into a fully packed chromosome.

TEI Testable Questions

Lesson Question: What Are the Processes Involved in Cellular Division?

`Recommended 10 minutes`

TEACHER NOTE Misconception: Students may think that organisms grow because their cells grow larger. In fact, although cells do grow, most growth exhibited by organisms is because of cell division. There is a limit to how large cells can become, so cell division allows organisms to grow while cells stay relatively small.

- Core Interactive Text: What Are the Processes Involved in Cellular Division?
- Image: Single-Celled Eukaryotic Cell Division

Lesson Question: What Are the Differences Between Cell Division Among Prokaryotes and Eukaryotes?

`Recommended 35 minutes`

- Core Interactive Text: What Are the Differences Between Cell Division Among Prokaryotes and Eukaryotes?
- Exploration: Exploring the Cell Cycle and Cell Division
- Image: Mitosis in Onion Root Tip

Lesson Question: What Is the Cell Cycle?

`Recommended 45 minutes`

- Core Interactive Text: What Is the Cell Cycle?
- Image: The Phases of the Cell Cycle
- Image: Diagram of Mitotic Cell Division
- Video: Interphase
- Video: Prophase
- Video: Prometaphase, Metaphase, Anaphase, Telophase, and Cytokinesis
- Video: Mitosis in Detail
- Video: Cytokinesis
- Video: Bone Regeneration

TEACHER NOTE Science and Engineering Practices: Developing and Using Models: In this item, students develop models of cell division. They develop and use multiple types of models to provide mechanistic accounts and predict phenomena. Help students describe the cell division process by using the Paper Slide strategy. As they learn about mitosis, have them draw the important events on a paper slide. Have teams sequence the events of mitosis in the correct order. The "Paper Slide" strategy is found on the Professional Learning tab. Click on Strategies & Resources, then click on Spotlight on Strategies (SOS). Now click on Sequence, then click on Spotlight on Strategies: Paper Slide.

TEI Mitosis

Lesson Question: What Are the Characteristics of Cancer Cells?   `Recommended 45 minutes`

TEACHER NOTE **Misconception:** Students may think that cancer results when healthy cells are directly damaged by an outside factor and the cells die. In fact, cancer is caused by the uncontrolled growth of once-healthy cells that have undergone multiple genetic changes. This uncontrolled growth impairs normal function in noncancerous cells.

- Core Interactive Text: What Are the Characteristics of Cancer Cells?
- Video: Cancer Cells
- Reading Passage: Careers in Cancer Research

TEACHER NOTE **Crosscutting Concepts: Stability and Change:** In this item, students analyze the change that occurs in cells during oncogenesis. They quantify and model changes in systems of cancer cells. Help students compare a cancer karyotype with a normal karyotype by placing the two karyotypes side by side on the board. Have students come up and circle each characteristic they observe that is different between the two cells.

TEI **Cancer Karyotype**

Explore More Resources

Resources in Explore More Resources support differentiation within your classroom by

- providing additional visualization of content
- affording extension of content to those students ready for acceleration
- offering Lexile reading levels for reading passages

Online explorations and hands-on experiences are provided so that students can conduct virtual investigations, collect and design investigations, and collect and analyze data; these skills are essential to developing scientific understanding.

Explain `45–90 minutes`

In Explore, students
1. uncovered scientific understandings
2. conducted investigations
3. analyzed data, text, and other media resources
4. collected evidence to support their scientific explanation

In Explain, provide students with time to formally compose their scientific explanations around the Explain or student-generated questions using evidence collected from Explore.

Scientific explanations are student responses, either written or orally presented, that explain scientific phenomena based upon evidence. Developing a scientific explanation requires students to analyze and interpret data to construct meaning out of the data. There are three main components to the scientific explanation: the claim, the evidence, and the reasoning.

To help students to communicate their scientific explanations, allow them to utilize the multimedia creation tools such as Board Builder and Whiteboard. Remind them that they may upload image, audio, and video files using the "attach file" optionto communicate their scientific explanations.

Students may construct their scientific explanations individually or within a small group of students. Students should communicate their explanations with other classmates, and provide constructive criticism and refine their explanations prior to submission to the teacher. If explanations are used as a formative assessment, you can provide additional feedback and comments to support students as they refine their explanations.

EXPLAIN?

Explain what steps are involved in the cell division of a prokaryote versus a eukaryote.

Elaborate with STEM   45–135 minutes

*Elaborate with STEM are optional extension resources available after students have demonstrated proficiency with standards addressed previously in the concept.

NGSS Components

SEP	CCC
■ Asking Questions and Defining Problems ■ Planning and Carrying Out Investigation ■ Using Mathematics and Computational Thinking ■ Constructing Explanations and DesigningSolutions ■ Obtaining, Evaluating, and Communicating Information	■ Cause and Effect ■ Structure and Function

STEM In Action   45 minutes

STEM in Action ties the scientific concepts to real-world applications, with many connecting to STEM careers. Technology Enhanced Items (TEIs) expect students to critically read the Core Interactive Text (CIT) and review the provided media resources.

Applying Cell Division

- Core Interactive Text: Applying Cell Division
- Image: Man Receiving Radiation Therapy
- Video: Breakthrough in Early Cancer Detection
- Video: Developing New Cures Using Stem Cells
- Video: Cell Differentiation

TEI Speeding or Slowing Down?

TEI Speeding or Slowing Down? Pt. 2

STEM Project Starters

STEM Project Starters provide additional real-world contexts that require students to apply and extend their content knowledge related to the concept. STEM Project Starters can also serve as an alternative instructional hook presented at the beginning of the learning progression. The project can then be revisited throughout and at the end of the 5E learning cycle, for students to apply content knowledge.

STEM Project Starter Environmental Effects on the Mitosis of Onions

Recommended 90 minutes

What environmental substances affect the rate of mitosis on onions?

TEACHER NOTE Students should submit their reports at least twice—once when they have designed their procedure and once when they have conducted the investigation. Use this as a formative assessment to assess student understanding on how various factors affect the rate of cell division.

TEI Onion Investigation

STEM Project Starter: Technology and Stem Cells

Recommended 60 minutes

How are stem cells currently used?

TEACHER NOTE You might want to provide a list of acceptable topics and uses before assigning this project. You may also choose to have students create a presentation on their topic. Use this as a formative assessment to gauge student understanding of the implications of stem cells.

TEI Technology and Stem Cells

■ Activity Engineering Design Sheet

Evaluate `45–90 minutes`

Explain Question:
Explain what steps are involved in the cell division of a prokaryote versus a eukaryote.

Lesson Questions (LQ):
- What is the structure of chromosomes?
- What are the processes involved in cellular division?
- What are the differences between cell division among prokaryotes and eukaryotes?
- What is the cell cycle?
- What are the characteristics of cancer cells?

Throughout instruction and the 5E learning cycle, you will have collected formative assessment data to drive the assignment of resources and experiences to students. Evaluate is intended to include summative assessment checks for proficiency. You can use the Explain and Lesson Questions for the concept as a summative assessment in a variety of ways such as these:

- Post each Lesson Question (LQ) in various locations in the classroom, and have small groups of students generate claim statements related to the Lesson Question (LQ). Other students can add to the claim, or refute the claim, during a gallery walk where they place additional pieces of evidence on each Lesson Question (LQ) poster.
- Assign small groups of students to each Lesson Question (LQ) and have the groups generate a poster, board, graphic, or piece of text that answers the question. Use a jigsaw approach and create a secondset of groups that contain members from each Lesson Question (LQ) group to share their ideas.
- Ask students to return to their initial ideas for the Explain question and add additional details and evidence.

Encourage students toreview the concept review and complete the Student Self-Check practice assessment prior to assigning the Summative Teacher Concept assessment.

- Student Review and Practice Assessment
- Teacher Concept Assessments

Asexual and Sexual Reproduction

The Five E Instructional Model

Science Techbook follows the 5E instructional model. This Model Lesson includes strategies for each of the 5Es. As you design the inquiry-based learning experience for students, be sure to collect data during instruction to drive your instructional decisions. Point-of-use teacher notes are also provided within each E-tab.

Engage 45–90 minutes

Engage Media Resources

The resources found in Engage are intended to stimulate students by exposing them to a phenomenon relevant to the content of the lesson. Engage also provides examples of relevant real-world applications that allow students to begin to make observations and relate the science content to their everyday lives. The Core Interactive Text (CIT) and media resources are carefully designed to prompt students to begin asking questions that they can investigate during the Explore phase of the lesson. They should also start collecting evidence to address the Explain question located at the bottom of the Engage page.

> **TEACHER NOTE** **Investigative Phenomenon:** Hardly anyone can resist the appeal of babies. Many baby animals, like these puppies, start out as small, young versions of their parents. Use the picture and the following questions to stimulate student thinking about reproduction. How did these puppies come into being? What did they grow from? What controlled their growth to ensure they turned out as puppies and not, say, kittens? Why do they look like miniature versions of adult dogs? Do the young of organisms all look like their parents? If not, why not? And if they do, are they identical to one of their parents?

- Core Interactive Text: Exploring Asexual and Sexual Reproduction
- Video: Puppies!
- Video: Reproduction
- Image: Binary Fission
- Image: Lion Family

Explain Question

The Explain question focuses students on gathering information in the Explore section. The Explain question can be used to

- Record what students already know related to the Explain question.
- Serve as a template or model for students to generate their own scientific questions.
- Collect evidence as students work through the lesson.
- Allow students to reflect on their growth before and after the lesson.

Explain

Explain what happens during asexual and sexual reproduction.

- Image: A Little Help from Your Friends

Engage Formative Assessment

Technology Enhanced Items (TEIs) found on the Engage page enable you to collect data on students' prior knowledge and identify the common misconceptions they may possess that are related to the topic of study. These items are designed as quick checks for understanding and allow each student one attempt at each question. You can use the data collected to decide whether to assign additional resources to the class, or determine what individual or groups of students may need reinforcement or accelerated learning, prior to completing the Explore portion of the lesson.

TEACHER NOTE Purpose: Use this student response to evaluate students' prior knowledge of the concept. The Model Lesson provides information on common student misconceptions. Suggested use: class discussion.

TEI Explaining Asexual Reproduction

Before You Begin

What Do You Already Know about Asexual and Sexual Reproduction?

TEACHER NOTE Purpose: a formative assessment used to evaluate prior knowledge and misconceptions related to asexual and sexual reproduction. Suggested use: individual work.

TEI Asexual or Sexual?
TEI Steps in Sexual Reproduction

TEACHER NOTE Purpose: a formative assessment used to evaluate prior knowledge and misconceptions related to asexual and sexual reproduction. Suggested use: individual work.

TEI Comparing Meiosis and Mitosis

- Video: Genes: The Blueprint of Life
- Video: Review of DNA Replication
- Video: Review of Mitosis

Explore `315 minutes`

Lesson Questions (LQs):

1. What are the differences between asexual and sexual reproduction?
2. What are haploid and diploid cells?
3. What is the function of each stage of meiosis?
4. What are the differences between mitosis and meiosis?
5. What are spermatogenesis and oogenesis?
6. What are fertilization and development?
7. What is Somatic Cell Nuclear Transfer (SCNT)?
8. What are the effects of non-disjunction?
9. What are some examples of aneuploidy, monosomy, and trisomy?

Effective science instruction involves a student-centered rather than a teacher-centered approach. This can be accomplished either with Directed Inquiry or Guided Inquiry, depending on the needs and abilities of your class. Encourage students to select a variety of resources in their pursuit of answers as they work through Explore, with the end goal of constructing their scientific explanation in the Explain tab.

Directed Inquiry	Guided Inquiry
In Directed Inquiry, teachers provide students with a sequence of specified resources, challenging questions, and clear outcomes. Within this context students are given the opportunity to interact independently with each resource as prescribed by the teacher. Often different students groups can be guided through several different resources at the same time. For example, one group could work on a reading passage while a second group conducts a small-group Hands-On Activity with the teacher, and a third group is independently engaged with an online interactive resource.	In Guided Inquiry, students have independence to decide the scope and sequence of their investigations. Using resources from Techbook, students determine for themselves which resources they will Explore to answer the Lesson Questions. It is important to note that each student will choose multiple resources, but no one student is expected to use all the resources available. Students also determine the order in which to explore these resources and how to record their findings.

NGSS Components

SEP	CCC
■ Analyzing and Interpreting Data ■ Constructing Explanations and Designing Solutions	■ Patterns ■ Structure and Function ■ Stability and Change

Lesson Question: What Are the Differences Between Asexual and Sexual Reproduction?

Recommended 55 minutes

TEACHER NOTE Connections: Crosscutting Concept: Structure and Function: As students engage with each of the lesson questions in this concept, they will build on their understanding of the cell as a system.

Students will examine the properties of the parts of the cell involved in reproduction, and their structures and interconnections, to reveal the function of the reproductive system in both sexual and asexual organisms. They will deduce the functions and properties of reproductive structures from their overall structure, the way their components are shaped and used, and the molecular substructures of their various materials.

TEACHER NOTE Misconception: Students may think that asexual reproduction is restricted to microorganisms only. Although it is the most common form of reproduction in microorganisms, other living things also reproduce asexually. Regeneration, parthenogenesis, and internal budding are all forms of asexual reproduction that larger life forms undergo. For example, echinoderms and many plant species regenerate, some insects such as bees and wasps undergo parthenogenesis, and sponges bud internally.

- Core Interactive Text: What Are the Differences Between Asexual and Sexual Reproduction?
- Image: Budding
- Video: Characteristics of Prokaryotes
- Video: Types of Reproduction

TEACHER NOTE Misconception: Students may believe sexual reproduction always involves mating. They may focus on mammalian reproduction. However, sexual reproduction only requires the fusion of male and female gametes. This can occur in several different ways other than copulation, including pollination, amplexus, and other forms of external fertilization. In addition, students may not understand that plants can sexually reproduce. Since they do not copulate, it might not seem obvious that plants reproduce sexually, but pollination is one component of sexual reproduction in some plants.

Formative Assessment:

Throughout Explore, Technology Enhanced Items (TEIs) are embedded as multi-dimensional formative checks for understanding. You can use the data they provide to

- assign additional support
- extend learning
- design additional learning tasks to clarify student misconceptions

The Explore TEIs provide students with three attempts to demonstrate their proficiency. Scaffolded feedback is provided for each attempt. If a student does not achieve proficiency by the third attempt, a media asset is provided as an additional learning opportunity.

TEACHER NOTE Practices: Science and Engineering Practice: Constructing Explanations and Designing Solutions: This item allows students to apply scientific reasoning, theory, and/or models to link evidence to claims about sexual and asexual reproduction to assess the extent to which the reasoning and data support the explanation regarding their similarities and differences. Extend this item by having students create a two-column chart with headers "Asexual Reproduction" and "Sexual Reproduction" into which they add descriptions of organisms according to their mode of reproduction. This can be an ongoing project throughout the remainder of the lesson, giving students the opportunity to research and obtain information about different types of organisms and communicate this information via their charts.

TEI Comparing Asexual and Sexual Reproduction

Lesson Question: What Are Haploid and Diploid Cells? **Recommended 25 minutes**

- Core Interactive Text: What Are Haploid and Diploid Cells?
- Image: Sperm and Egg Cells
- Image: Diploid and Haploid Cells

Lesson Question: What Is the Function of Each Stage of Meiosis? **Recommended 45 minutes**

- Core Interactive Text: What Is the Function of Each Stage of Meiosis?
- Exploration: Asexual and Sexual Reproduction
- Video: Phases of Meiosis: Part One
- Video: Phases of Meiosis: Part Two
- Image: Meiosis I and II
- Video: Meiosis, Germ Cells, and Interphase
- Video: Meiosis

TEACHER NOTE Connections: Crosscutting Concept: Stability and Change: These items engage students in actively processing the changes taking place in cells during meiosis. These items are good assessments to assign just after students have completed the exploration, "Exploring Asexual and Sexual Reproduction." In this item, students understand much of science deals with constructing explanations of how things change and how they remain stable. They quantify and model changes in reproductive systems over very short or very long periods of time. They see some changes are irreversible, and negative feedback can stabilize a system, while positive feedback can destabilize it. They recognize systems can be designed for greater or lesser stability. As a follow-up activity, ask students to write explanatory sentences to accompany the flow chart they completed in the second item. The sentences should explain the basis for the change at each level of the flow chart.

TEI Cell Stages of Meiosis

TEI Haploid and Diploid Stages

Lesson Question: What Are the Differences Between Mitosis and Meiosis?

`Recommended 30 minutes`

- Core Interactive Text: What Are the Differences Between Mitosis and Meiosis?
- Video: Mitosis and Meiosis

TEACHER NOTE **Practices: Science and Engineering Practice: Analyzing and Interpreting Data:** This item requires students to analyze and interpret data presented in graph form using tools, technologies, and/or models (e.g., computational, mathematical) to make valid and reliable scientific claims about changes taking place during mitosis and changes taking place during meiosis. Extend this item by having students draw cell models showing changes in chromosomes at different stages that coincide with the changes in DNA mass on each graph.

TEI Mitosis and Meiosis Comparison

Lesson Question: What Are Spermatogenesis and Oogenesis?

`Recommended 35 minutes`

- Core Interactive Text: What Are Spermatogenesis and Oogenesis?
- Image: Spermatogenesis
- Video: What Is Spermatogenesis?
- Image: Oogenesis
- Video: What Is Oogenesis?

Lesson Question: What Are Fertilization and Development?

`Recommended 45 minutes`

- Core Interactive Text: What Are Fertilization and Development?
- Image: The Moment of Fertilization
- Video: What Happens During Fertilization?
- Image: Gastrulation
- Video: Cellular Differentiation in the Embryo
- Video: Differentiation and the Fate of Cells
- Image: Early Development

Lesson Question: What Is Somatic Cell Nuclear Transfer (SCNT)?

`Recommended 20 minutes`

- Core Interactive Text: What Is Somatic Cell Nuclear Transfer (SCNT)?
- Video: Somatic Cell Nuclear Transfer
- Video: SCNT: Controversy or Opportunity?

TEACHER NOTE Connections: Crosscutting Concept: Patterns: In completing this item students observe patterns in reproductive systems at different scales (both molecular and organismal) and cite patterns as empirical evidence for causality in supporting their explanations of the type of reproductive technology used to produce three calves. You may wish to have a short class discussion before assigning this item to talk about the similarities and differences between SCNT and In Vitro Fertilization (IVF) techniques. After students have completed this item, extend student learning by having them draw sketches to show steps in these two processes. Ask students to use their drawings to justify the answers they chose in this assessment item.

TEI Applying New Reproductive Technologies

Lesson Question: What Are the Effects of Non-Disjunction?   *Recommended 25 minutes*

- Core Interactive Text: What Are the Effects of Non-Disjunction?
- Image: Non-Disjunction
- Video: Non-Disjunction

Lesson Question: What Are Some Examples of Aneuploidy, Monosomy, and Trisomy?   *Recommended 35 minutes*

- Core Interactive Text: What Are Some Examples of Aneuploidy, Monosomy, and Trisomy?
- Image: Karyotype of Monosomy
- Video: Aneuploidy
- Video: Aneuploidy: Monosomy and Trisomy

TEACHER NOTE Practices: Science and Engineering Practice: Analyzing and Interpreting Data: In this item, students apply what they have learned in the last two Explore sections as they analyze and interpret data using tools, technologies, and/or models (e.g., computational, mathematical) to make valid and reliable scientific claims about two human karyotypes. Extend this item by having students draw sketches to show how non-disjunction results in each of the two chromosomal abnormalities represented by the two karyotypes.

TEI Analyzing Karyotypes

Explore More Resources

Resources in Explore More Resources support differentiation within your classroom by

- providing additional visualization of content
- affording extension of content to those students ready for acceleration
- offering Lexile reading levels for reading passages

Online explorations and hands-on experiences are provided so that students can conduct virtual investigations, collect and design investigations, and collect and analyze data; these skills are essential to developing scientific understanding.

Explain 45–90 minutes

In Explore, students
1. uncovered scientific understandings
2. conducted investigations
3. analyzed data, text, and other media resources
4. collected evidence to support their scientific explanation

In Explain, provide students with time to formally compose their scientific explanations around the Explain or student-generated questions using evidence collected from Explore.

Scientific explanations are student responses, either written or orally presented, that explain scientific phenomena based upon evidence. Developing a scientific explanation requires students to analyze and interpret data to construct meaning out of the data. There are three main components to the scientific explanation: the claim, the evidence, and the reasoning.

To help students to communicate their scientific explanations, allow them to utilize the multimedia creation tools such as Board Builder and Whiteboard. Remind them that they may upload image, audio, and video files using the "attach file" option to communicate their scientific explanations.

Students may construct their scientific explanations individually or within a small group of students. Students should communicate their explanations with other classmates, and provide constructive criticism and refine their explanations prior to submission to the teacher. If explanations are used as a formative assessment, you can provide additional feedback and comments to support students as they refine their explanations.

CAN YOU EXPLAIN?
Explain what happens during asexual and sexual reproduction.

Elaborate with STEM `45–135 minutes`

*Elaborate with STEM are optional extension resources available after students have demonstrated proficiency with standards addressed previously in the concept.

NGSS Components

SEP	CCC
■ Analyzing and Interpreting Data ■ Engaging in Argument from Evidence ■ Obtaining, Evaluating, and Communicating Information	■ Systems and System Models ■ Stability and Change

STEM In Action `45 minutes`

STEM in Action ties the scientific concepts to real-world applications, with many connecting to STEM careers. Technology Enhanced Items (TEIs) expect students to critically read the Core Interactive Text (CIT) and review the provided media resources.

Applying Asexual and Sexual Reproduction

- Core Interactive Text: Applying Asexual and Sexual Reproduction
- Image: IVF
- Video: An In Vitro Experiment
- Video: Cloning the Tasmanian Tiger
- Video: Cloning Human Beings
- Video: Genetic Engineering

TEACHER NOTE Purpose: to evaluate student ability to interpret graphs and to explain concepts learned in lesson. Suggested use: think-pair-share then write response.

TEI In Vitro Fertilization

STEM Project Starters

STEM Project Starters provide additional real-world contexts that require students to apply and extend their content knowledge related to the concept. STEM Project Starters can also serve as an alternative instructional hook presented at the beginning of the learning progression. The project can then be revisited throughout and at the end of the 5E learning cycle, for students to apply content knowledge.

STEM Project Starter Ethics of Cloning

> Recommended 45 minutes

Why is cloning controversial?

TEACHER NOTE Purpose: to evaluate student understanding of the process of SCNT technology and its ethical implications. Suggested use: class discussion and debate.

TEI Ethics of Cloning

STEM Project Starter Meiosis Math

> Recommended 90 minutes

Can you tell what phase of meiosis is shown here?

TEACHER NOTE Purpose: a formative assessment used to test student understanding of the differences of meiosis, by the numbers. Suggested use: individual work.

■ Hands-On Lab: A Cycle Through Reproduction

TEI Meiosis Math

Evaluate `45–90 minutes`

Explain Question:
Explain what happens during asexual and sexual reproduction

Lesson Questions (LQ):
- What are the differences between asexual and sexual reproduction?
- What are haploid and diploid cells?
- What is the function of each stage of meiosis?
- What are the differences between mitosis and meiosis?
- What are spermatogenesis and oogenesis?
- What are fertilization and development?
- What is Somatic Cell Nuclear Transfer (SCNT)?
- What are the effects of non-disjunction?
- What are some examples of aneuploidy, monosomy, and trisomy?

Throughout instruction and the 5E learning cycle, you will have collected formative assessment data to drive the assignment of resources and experiences to students. Evaluate is intended to include summative assessment checks for proficiency. You can use the Explain and Lesson Questions for the concept as a summative assessment in a variety of ways such as these:

- Post each Lesson Question (LQ) in various locations in the classroom, and have small groups of students generate claim statements related to the Lesson Question (LQ). Other students can add to the claim, or refute the claim, during a gallery walk where they place additional pieces of evidence on eachLesson Question (LQ) poster.
- Assign small groups of students to each Lesson Question (LQ) and have the groups generate a poster,board, graphic, or piece of text that answers the question. Use a jigsaw approach and create a secondset of groups that contain members from each Lesson Question (LQ) group to share their ideas.
- Ask students to return to their initial ideas for the Explain question and add additional details and evidence.

Encourage students to review the concept review and complete the Student Self-Check practice assessment prior to assigning the Summative Teacher Concept assessment.

- Student Review and Practice Assessment
- Teacher Concept Assessments

Biological Organization and Control

The Five E Instructional Model

Science Techbook follows the 5E instructional model. This Model Lesson includes strategies for each of the 5Es. As you design the inquiry-based learning experience for students, be sure to collect data during instruction to drive your instructional decisions. Point-of-use teacher notes are also provided within each E-tab.

Engage   45–90 minutes

Engage Media Resources

The resources found in Engage are intended to stimulate students by exposing them to a phenomenon relevant to the content of the lesson. Engage also provides examples of relevant real-world applications that allow students to begin to make observations and relate the science content to their everyday lives. The Core Interactive Text (CIT) and media resources are carefully designed to prompt students to begin asking questions that they can investigate during the Explore phase of the lesson. They should also start collecting evidence to address the Explain question located at the bottom of the Engage page.

> **TEACHER NOTE** **Investigative Phenomena:** This lesson uses exercising as a way of introducing how different parts of an organism work together to perform the variety of functions required for an athlete to run. The scenario has as its main focus the cardiovascular system but references to other body processes, systems, and organs that students will encounter as they work through the lesson. Students will most likely be familiar with many of the body systems mentioned in this lesson. As they work through Engage, have them try to identify all the systems that are involved in the scenario.

- Core Interactive Text: Working Together
- Image: Run Your Way to Better Fitness
- Image: Exercise and Cardiovascular Activity

Explain Question

The Explain question focuses students on gathering information in the Explore section. The Explain question can be used to

- Record what students already know related to the Explain question.
- Serve as a template or model for students to generate their own scientific questions.
- Collect evidence as students work through the lesson.
- Allow students to reflect on their growth before and after the lesson.

Explain

How do the different systems of the body work together to maintain internal conditions?

- Image Jumping Hurdles

Engage Formative Assessment

Technology Enhanced Items (TEIs) found on the Engage page enable you to collect data on students' prior knowledge and identify the common misconceptions they may possess that are related to the topic of study. These items are designed as quick checks for understanding and allow each student one attempt at each question. You can use the data collected to decide whether to assign additional resources to the class, or determine what individual or groups of students may need reinforcement or accelerated learning, prior to completing the Explore portion of the lesson.

TEACHER NOTE Use student responses to this formative assessment to evaluate their basic knowledge of how body systems are organized to maintain internal conditions. Suggested use of this item includes students working in think-pair-shares to discuss their thoughts about the role of biological organization in maintaining internal conditions.

TEI Some Like It Hot

Before You Begin

What Do I already know about biological organization and control?

TEACHER NOTE This formative assessment is intended to assess student misconceptions about specialized cells. Students might think that structure is consistent across all organisms or be unclear on the differing levels of organization in living things. Students can discuss their answers with a think-pair-share partner.

TEI Cellular Structures

TEACHER NOTE This formative assessment item uses the example of the muscular system to provide the teacher with feedback on prior knowledge about muscle structure and function. In middle school, students should have learned that organs such as specific muscles are made of cells and interact with the parts of other organ systems and how these systems may overlap in functions. Present this review to find out where gaps in knowledge exist.

TEI Body Systems

TEACHER NOTE Use student responses to this item to provide feedback on whether students can associate the primary function of an organ system with its role in homeostasis. It also addresses misconceptions that students have about the functions of the body's various systems. This activity should be completed individually, followed by class review.

TEI Organ Systems

- Video: Respiration
- Video: Sweaty Boots
- Video: Maximizing Power Output

Explore `270 minutes`

Lesson Questions (LQs):

1. How are organisms organized into cells, tissues, and organs?
2. How do organs work together to form specific functions?
3. Why and how do organisms maintain their internal state?
4. How do different parts of the body interact to maintain homeostasis?

Effective science instruction involves a student-centered rather than a teacher-centered approach. This can be accomplished either with Directed Inquiry or Guided Inquiry, depending on the needs and abilities of your class. Encourage students to select a variety of resources in their pursuit of answers as they work through Explore, with the end goal of constructing their scientific explanation in the Explain tab.

Directed Inquiry	Guided Inquiry
In Directed Inquiry, teachers provide students with a sequence of specified resources, challenging questions, and clear outcomes. Within this context students are given the opportunity to interact independently with each resource as prescribed by the teacher. Often different students groups can be guided through several different resources at the same time. For example, one group could work on a reading passage while a second group conducts a small-group Hands-On Activity with the teacher, and a third group is independently engaged with an online interactive resource.	In Guided Inquiry, students have independence to decide the scope and sequence of their investigations. Using resources from Tech book, students determine for themselves which resources they will Explore to answer the Lesson Questions. It is important to note that each student will choose multiple resources, but no one student is expected to use all the resources available. Students also determine the order in which to explore these resources and how to record their findings.

NGSS Components

SEP	CCC
■ Obtaining, Evaluating, and Communicating Information ■ Analyzing and Interpreting Data ■ Developing and Using Models ■ Engaging in Argument from Evidence ■ Planning and Carrying Out investigations ■ Scientific Investigations Use a Variety of Methods	■ Cause and Effect ■ Stability and Change ■ Structure and Function ■ Systems and System Models

Lesson Question: How are organisms organized into cells, tissues, and organs?

> Recommended 90 minutes

TEACHER NOTE Connections: Crosscutting Concept: Scale, Proportion and Quantity: Students should be familiar with basic cell structure and function. This lesson question revisits some of the material students have encountered in earlier concepts about cell structure, function and division. The emphasis in this first lesson question is on how the smallest structures within cells, cells, and the structures—tissues and organs—into which they are organized, function at different scales, proportions and quantities.

As students read and comprehend complex texts, view the videos, and complete the interactives, labs, and other Hands-On Activities, have them summarize and obtain scientific and technical information. Students will use this evidence to support their initial ideas on how to answer the Explain Question or their own question they generated during Engage. Have students record their evidence using "My Notebook."

- Core Interactive Text: How Are Organisms Organized into Cells, Tissues, and Organs?
- Video: Cells in the Human Body
- Video: Types of Human Cells
- Video: Function of Muscles
- Image: Three Types of Muscle
- Video: The Different Types of Muscle
- Video: The Mechanism of Muscle Fiber Contraction
- Activity: Hands-On Lab: Skeletal Muscles
- Video: Neurons
- Image: Vascular Bundle
- Video: Plant Stems
- Video: Fine Detail of Xylem Structure

Formative Assessment:

Throughout Explore, Technology Enhanced Items (TEIs) are embedded as multi-dimensional formative checks for understanding. You can use the data they provide to

- assign additional support
- extend learning
- design additional learning tasks to clarify student misconceptions

The Explore TEIs provide students with three attempts to demonstrate their proficiency. Scaffolded feedback is provided for each attempt. If a student does not achieve proficiency by the third attempt, a media asset is provided as an additional learning opportunity.

TEI Organelles, Cells, Tissues, and Organs

Lesson Question How Do Organs Work Together
to Form Specific Functions?

Recommended 45 minutes

- Core Interactive Text: How Do Organs Work Together to Form Specific Functions?

TEACHER NOTE Connections: Crosscutting Concepts: Systems and System Models: Have students work in small groups to review the table below. Each group should create a concept map using arrows labeled with the relationships between the body systems. Have students speculate what would happen if one of the systems or system functions were disrupted.

- Table: Systems of the Human Body

TEACHER NOTE Science and Engineering Practices: Obtaining, Evaluating, and Communicating Information: Have students select an organ system and use the videos below and in Explore More Resources (and other resources where available) to further research its function and the role of its major organs. Students should produce a board or poster outlining the results of their research and share it with other members of the class.

- Video: The Heart
- Video: The Digestive System
- Video: The Integumentary System
- Video: The Immune System
- Video: Excretory System
- Video: What Systems Keep the Body Running

TEACHER NOTE Connections: Crosscutting Concepts: Structure and Function: Students should use the exploration of The Digestive System to learn more about the digestive organs and their functions. Once they have completed the exploration, they should work to complete the item "Indigestion."

- Exploration: The Digestive System

TEI Indigestion

Lesson Question: Why and How Do Organisms Maintain
Their Internal State?

Recommended 45 minutes

- Core Interactive Text: Why and How Do Organisms Maintain Their Internal State?
- Video: Endotherms: Body Temperature Under Control
- Video: What is Homeostasis?

TEACHER NOTE Misconception: Students may think that the body's internal conditions are stable and constant. In fact, as internal conditions change, body systems respond to the changing conditions to maintain internal equilibrium. Body systems constantly monitor internal and external conditions and respond to return internal conditions to an equilibrium state.

TEACHER NOTE Connections: Crosscutting Concepts: Systems and System Models: In this section of the lesson question, students must relate the working of a simple mechanical device, a thermostat, to a more complex control mechanism in an organism. Spend time going through the diagram of the negative feedback loop for this model, so they have a good grasp of how negative feedback works. Have students work in pair or small groups to identify and explain other examples of such mechanisms that they have encountered (such as toilet cisterns and computer cooling systems).

- Image: Thermostat
- Image: Negative Feedback Loop

TEACHER NOTE Misconception: Students may be aware that animals use homeostasis to control their internal environment but have not considered homeostasis in microorganisms or plants. Refer to students existing knowledge about osmoregulation—in plant cells and amoeba—and use this to emphasize that all organisms must regulate their internal environment to conduct their necessary life processes. The item "How Can Plants Use Water More Efficiently?" provides an additional example of non-animal homeostasis.

- Video: Negative Feedback Loops
- Video: The Structure and Function of Stomata

TEACHER NOTE Science and Engineering Practices: Analyzing and Interpreting Data: Have students view the video "The Structure and Function of Stomata." After the video, have students summarize it by writing a six-word story. This strategy can be found in the professional development tab, on the Spotlight on Strategies page under summarizing.

Next, have students work on the item below in which students analyze data from a scientific investigation that studied the effects of modifying plants to influence the stomata, a major structure in maintaining plant homeostasis. They will use the data to make a valid conclusion about the study. Stomata respond to environmental conditions to maintain homeostasis.

TEI How Can Plant Use Water More Efficiently?

Lesson Question How Do Different Parts of the Body Interact to Maintain Homeostasis?

Recommended 90 minutes

- Core Interactive Text: How Do Different Parts of the Body Interact to Maintain Homeostasis?

TEACHER NOTE **Science and Engineering Practices: Developing and Using Models:** The following sections provide examples of some of the homeostatic processes occurring in the human body. After reviewing the material below, have groups select an example and create a flow diagram that models one of these homeostatic processes/feedback loops (or another of their choosing). Have groups share and critique each other's flow diagrams.

TEACHER NOTE **Misconception:** Students may think that shivering and sweating are simply indicators of temperatures of the external environment. In fact, shivering and sweating serve to raise or lower the internal temperature of the human body.

- Image: Basking
- Video: Thermoregulation
- Exploration: Homeostasis
- Video: The Excretory System
- Video: The Flow of Blood
- Image: Control of Blood Pressure
- Hands-On Activity: Homeostasis and Heart Rate

TEACHER NOTE **Practices: Science and Engineering Practices: Engaging in Argument from Evidence:** In this item, the student is asked to make a claim based on evidence about human systems. They will classify biological processes as positive or negative feedback loops. Students must make their choices based on valid and reliable evidence, scientific reasoning, and principles of homeostasis obtained during this lesson. Comparing negative and positive feedback systems will enhance the students' understanding of homeostasis and the various methods of achieving balance in our bodies. Extend this item by having students represent positive and negative feedback loops on a line graph. Then compare the two mechanisms.

TEI **Positive or Negative Feedback Loop**

Explore More Resources

Resources in Explore More Resources support differentiation within your classroom by

- providing additional visualization of content
- affording extension of content to those students ready for acceleration
- offering Lexile reading levels for reading passages

Online explorations and hands-on experiences are provided so that students can conduct virtual investigations, collect and design investigations, and collect and analyze data; these skills are essential to developing scientific understanding.

Explain `45–90 minutes`

In Explore, students
1. uncovered scientific understandings
2. conducted investigations
3. analyzed data, text, and other media resources
4. collected evidence to support their scientific explanation

In Explain, provide students with time to formally compose their scientific explanations around the Explain or student-generated questions using evidence collected from Explore.

Scientific explanations are student responses, either written or orally presented, that explain scientific phenomena based upon evidence. Developing a scientific explanation requires students to analyze and interpret data to construct meaning out of the data. There are three main components to the scientific explanation: the claim, the evidence, and the reasoning.

To help students to communicate their scientific explanations, allow them to utilize the multimedia creation tools such as Studio and Whiteboard. Remind them that they may upload image, audio, and video files using the "attach file" option to communicate their scientific explanations.

Students may construct their scientific explanations individually or within a small group of students. Students should communicate their explanations with other classmates, and provide constructive criticism and refine their explanations prior to submission to the teacher. If explanations are used as a formative assessment, you can provide additional feedback and comments to support students as they refine their explanations.

EXPLAIN

How do the different systems of the body work together to maintain internal conditions?

Elaborate with STEM `45–135 minutes`

*Elaborate with STEM are optional extension resources available after students have demonstrated proficiency with standards addressed previously in the concept.

NGSS Components

SEP	CCC
■ Asking Questions and Defining Problems ■ Engaging in Argument from Evidence ■ Using Mathematics and Computational Thinking ■ Constructing Explanations and Designing Solutions ■ Analyzing and Interpreting Data ■ Obtaining, Evaluating and Communicating Information	■ Cause and Effect ■ Structure and Function

STEM In Action `45 minutes`

STEM in Action ties the scientific concepts to real-world applications, with many connecting to STEM careers. Technology Enhanced Items (TEIs) expect students to critically read the Core Interactive Text (CIT) and review the provided media resources.

Applying Biological Organization and Control

- Core Interactive Text: Applying Biological Organization and Control
- Image: Emphysema Warning
- Image: Insulin Pump for Diabetics
- Video: Breathalyzer for Disease
- Video: Type I and Type II Diabetes
- Image The Story of Insulin

TEACHER NOTE Use student responses on this formative item to evaluate student understanding of feedback mechanisms and assess their ability to apply what they learned to blood glucose homeostasis. This could be considered as a think-pair-share activity.

TEI Blood Glucose Control

STEM Project Starters

STEM Project Starters provide additional real-world contexts that require students to apply and extend their content knowledge related to the concept. STEM Project Starters can also serve as an alternative instructional hook presented at the beginning of the learning progression. The project can then be revisited throughout and at the end of the 5E learning cycle, for students to apply content knowledge.

STEM Project Starter: Calcium Control

Recommended 45 minutes

How can osteoporosis be avoided?

TEACHER NOTE Use this summative project to assess student understanding of the material in this concept. Students should complete this activity individually.

- Image: Blood Calcium Levels and Hormone Levels

TEI T-Scores

- Feedback Loop

STEM Project Starter Addictive Substances

Recommended 90 minutes

What are the effects of addictive substances on the human nervous system?

TEACHER NOTE The topic of addiction is, unfortunately, all too relevant to today's teenagers. When assigning this formative assessment, be aware of, and sensitive to, current local events with regard to the abuse of addictive drugs, such as recent arrests or overdoses. Students may have family members or friends with addictions to pain relievers or other addictive substances; some may be struggling themselves. Consider partnering with a health teacher for this project, and, depending on your community, informing parents about the project in advance.

Encourage students to complete some research before writing their objectives, as the data they locate may help determine their research focus. For the math component, ensure that students understand that mathematical models can show how medicines behave in the body. Mathematical models can predict the rate of metabolism of addictive drugs, and hence show how long chemicals remain in the body.

Encourage students to structure their presentations or reports to include an introduction, supporting sentences, and a conclusion. Presentations and reports should pay attention to logical and smooth transitions as they address all bulleted questions as well as the language that is used. Have students use professional language and tone when writing about and discussing drug abuse and addiction, rather than using informal language or slang.

Instruct students to form pairs and perform a peer review of one another's writing, in which students provide feedback to each other. Students can use this feedback to improve their transitions, language, style, and introductions and conclusions.

Have students cite their sources and assess the reliability, validity, and usefulness of each source that is used for their research. This is particularly important in potentially controversial areas such as drug abuse and addiction treatment, in which claims from all sides can be exaggerated or based on anecdotal evidence.

TEI Rational Function

TEI Titration Curves

TEI Estimating Values 1

TEI Estimating Values 2

TEI End-Behavior of a Rational Function 1

TEI End-Behavior of a Rational Function 2

STEM Project Starter Fight or Flight

Recommended 45 minutes

How is fight or flight triggered?

- Image: A Hunting Lioness

TEACHER NOTE This formative assessment is a STEM project relating to the endocrine system because it introduces the fight or flight mechanism used by many organisms to deal with danger. This response is regulated by the hormone epinephrine secreted by the adrenal glands. The graph shows depictions of what happens to hormone levels temporarily and if levels continue to increase.

TEI Fight or Flight

Evaluate `45–90 minutes`

Explain Question:
How do the different systems of the body work together to maintain internal conditions?

Lesson Questions (LQ):
- How are organisms organized into cells, tissues, and organs?
- How do organs work together to form specific functions?
- Why and how do organisms maintain their internal state?
- How do different parts of the body interact to maintain homeostasis?

- Throughout instruction and the 5E learning cycle, you will have collected formative assessment data to drive the assignment of resources and experiences to students. Evaluate is intended to include summative assessment checks for proficiency. You can use the Explain and Lesson Questions for the concept as a summative assessment in a variety of ways such as these:

- Post each Lesson Question (LQ) in various locations in the classroom, and have small groups of students generate claim statements related to the Lesson Question (LQ). Other students can add to the claim, or refute the claim, during a gallery walk where they place additional pieces of evidence on each Lesson Question (LQ) poster.
- Assign small groups of students to each Lesson Question (LQ) and have the groups generate a poster, board, graphic, or piece of text that answers the question. Use a jigsaw approach and create a second set of groups that contain members from each Lesson Question (LQ) group to share their ideas.
- Ask students to return to their initial ideas for the Explain question and add additional details and evidence.

Encourage students to review the concept review and complete the Student Self-Check practice assessment prior to assigning the Summative Teacher Concept assessment.

- Student Review and Practice Assessment
- Teacher Concept Assessments

Natural Resources

The Five E Instructional Model

Science Techbook follows the 5E instructional model. This Model Lesson includes strategies for each
of the 5Es. As you design the inquiry-based learning experience for students, be sure to collect data during instruction to drive your instructional decisions. Point-of-use teacher notes are also provided within each E-tab.

Engage   45–90 minutes

Engage Media Resources

The resources found in Engage are intended to stimulate students by exposing them to a phenomenon relevant to the content of the lesson. Engage also provides examples of relevant real-world applications that allow students to begin to make observations and relate the science content to their everyday lives. The Core Interactive Text (CIT) and media resources are carefully designed to prompt students to begin asking questions that they can investigate during the Explore phase of the lesson. They should also start collecting evidence to address the Explain question located at the bottom of the Engage page.

> **TEACHER NOTE** **Investigative Phenomenon:** How do humans use natural resources? Have students answer this question by focusing on how they use energy during the course of a normal day. They should work in small groups, listing the ways they use energy and where they think it comes from. Students may first think that some of the resources they use are electricity to light their bedroom, or fuel used in cars or buses. But encourage them to think in wider terms, such as the energy in their food. Use the video Global Energy Consumption to initiate their thinking. Have groups share their ideas with the class to create a class list. Next, narrow it down to a single, simple household item (such as a newspaper or a can of cat food). Give the same groups five minutes to develop a flow chart that shows the different natural resources they need to make to be able to transport this item to their homes. As students draw their charts, they will discover that the connections to various natural resources become very complex. Use a few examples from these charts to illustrate how the use of natural resources is an integral and ubiquitous part of all their lives.

- Core Interactive Text: Thinking about Natural Resources
- Video: Global Energy Consumption
- Image: A Café at Night
- Video: Ballooning American Energy Use
- Video: Addicted to Oil

Explain Question

The Explain question focuses students on gathering information in the Explore section. The Explain question can be used to

- Record what students already know related to the Explain question.
- Serve as a template or model for students to generate their own scientific questions.
- Collect evidence as students work through the lesson.
- Allow students to reflect on their growth before and after the lesson.

Explain

How do we use natural resources, and why is conservation of natural resources important?

- Image: Keeping Pace with Need

Engage Formative Assessment

Technology Enhanced Items (TEIs) found on the Engage page enable you to collect data on students' prior knowledge and identify the common misconceptions they may possess that are related to the topic of study. These items are designed as quick checks for understanding and allow each student one attempt at each question. You can use the data collected to decide whether to assign additional resources to the class, or determine what individual or groups of students may need reinforcement or accelerated learning, prior to completing the Explore portion of the lesson.

TEACHER NOTE Use this student response to evaluate students' prior knowledge of the concept. The Model Lesson provides information on common student misconceptions.

TEI Natural Resources

Before You Begin

What Do I Already Know about Natural Resources?

TEACHER NOTE This formative item is intended to provide the teacher with feedback on prior knowledge of this topic and misconceptions. Students who place coal, natural gas, or nuclear power in the unlimited category may not realize that just because there is a lot of a resource, that does not mean it can never be depleted. Students may discuss their answers with a partner before entering them.

TEI Limited Resources

TEACHER NOTE This formative item is intended to provide the teacher with feedback on prior knowledge of this topic and misconceptions. Students may not think of intangible resources such as the sun as natural resources.

Students should complete this task individually.

TEI Powering Earth's Systems

TEACHER NOTE This formative activity is intended to provide the teacher with feedback on prior knowledge of this topic. Its focus is to ensure students understand the wide array of resources that may be present in any single activity in their lives. Students should discuss the question as a class before entering their answers.

TEI What Makes That Happen?

- Video: Transfer of Energy
- Video: The Great Energy Machine
- Video: The Power behind Weather
- Video: Water Cycle on Earth

Explore `135 minutes`

Lesson Questions (LQs):

1. What Are Natural Resources and How Do We Use Them?

Effective science instruction involves a student-centered rather than a teacher-centered approach. This can be accomplished either with Directed Inquiry or Guided Inquiry, depending on the needs and abilities of your class. Encourage students to select a variety of resources in their pursuit of answers as they work through Explore, with the end goal of constructing their scientific explanation in the Explain tab.

Directed Inquiry	Guided Inquiry
In Directed Inquiry, teachers provide students with a sequence of specified resources, challenging questions, and clear outcomes. Within this context students are given the opportunity to interact independently with each resource as prescribed by the teacher. Often different students groups can be guided through several different resources at the same time. For example, one group could work on a reading passage while a second group conducts a small-group Hands-On Activity with the teacher, and a third group is independently engaged with an online interactive resource.	In Guided Inquiry, students have independence to decide the scope and sequence of their investigations. Using resources from Techbook, students determine for themselves which resources they will Explore to answer the Lesson Questions. It is important to note that each student will choose multiple resources, but no one student is expected to use all the resources available. Students also determine the order in which to explore these resources and how to record their findings.

NGSS Components

SEP	CCC
■ Analyzing and Interpreting Data ■ Engaging in Argument from Evidence	■ Cause and Effect

Lesson Question: What Are Natural Resources and How Do We Use Them?

`Recommended 135 minutes`

TEACHER NOTE Science and Engineering Practices: Engaging in Argument From Evidence: Throughout this concept, students learn of the various ways we depend on and use natural resources. Make this connection personal by having students complete a T-Chart of natural resources they use over the course of one or two days. Label one side of the chart *Renewable* and the other side *Nonrenewable*. Then, after students have many items in each column, have them develop a claim about the role of natural resources in their own lives. Partners can then explain their own claims and the evidence used to support that claim and also evaluate their partner's claim, evidence, and reasoning.

- Core Interactive Text: What Are Natural Resources and How Do We Use Them?
- Exploration: Natural Resources
- Hands-on Activity: Becoming Aware of Resource Use
- Video: Our Thirst for Fossil Fuels
- Video: Coal
- Video: The End of Easy Oil
- Video: The Rock Cycle

TEACHER NOTE Misconception: Students may think that the supply of all natural resources on Earth is too great to be depleted. In fact, most resources may be depleted if they are used more quickly than they are replenished; this is inherent in the concept of nonrenewable resource. Nonrenewable resources like fossil fuels and metal ores form over millions of years. The supplies are limited and in many cases dwindling due to heavy demand.

TEACHER NOTE Misconception: Students may think that only solid materials (e.g., wood, coal, metals) are natural resources. In fact, energy sources such as solar radiation are also natural resources, as are water (which consists of matter in liquid form) and wind (which consists of matter in gaseous form).

Formative Assessment:

Throughout Explore, Technology Enhanced Items (TEIs) are embedded as multi-dimensional formative checks for understanding. You can use the data they provide to

- assign additional support
- extend learning
- design additional learning tasks to clarify student misconceptions

The Explore TEIs provide students with three attempts to demonstrate their proficiency. Scaffolded feedback is provided for each attempt. If a student does not achieve proficiency by the third attempt, a media asset is provided as an additional learning opportunity.

TEACHER NOTE Science and Engineering Practices: Analyzing and Interpreting Data: The item Producers of Rare Earth Elements requires students to convert numerical data into percentages. Once students have created the pie chart, lead a short discussion to have students compare the two data sets in order to determine which representation is most helpful for an analysis of global trends. Extend this item and deepen the mathematics connection by having students make pie charts for all of the years on this graph and then analyze the trends they see and develop an explanation to explain those trends.

TEI Producers of Rare Earth Elements

- Image: Clear-Cut Logging
- Video: The Importance of Soil
- Video: Renewable Energy Resources
- Video: Biomass as a Renewable Energy Resource
- Reading Passage: The Next Vanishing Resource?
- Writing Prompt: Exploring Whether Different Natural Resources are Renewable or Not
- Reading Passage: Biofuel in the Woods
- Video: Biofuels
- Video: Biomass Energy
- Video: Wind Energy
- Video: Tidal and Geothermal Energy

TEACHER NOTE **Crosscutting Concepts: Cause and Effect:** There is a causal relationship between how resources are obtained and used and their potential side effects. Understanding this relationship is necessary in order to understand the limitations of different natural resources. In this item, Uses and Challenges, students are asked to classify challenges caused or exacerbated by using different natural resources. Extend this item by having students refer back to their natural resource T-Charts (see the first Teacher Note) and develop a list of challenges associated with their own, personal uses of natural resources.

TEI Uses and Challenges

Explore More Resources

Resources in Explore More Resources support differentiation within your classroom by

- providing additional visualization of content
- affording extension of content to those students ready for acceleration
- offering Lexile reading levels for reading passages

Online explorations and hands-on experiences are provided so that students can conduct virtual investigations, collect and design investigations, and collect and analyze data; these skills are essential to developing scientific understanding.

Explain `45–90 minutes`

In Explore, students
1. uncovered scientific understandings
2. conducted investigations
3. analyzed data, text, and other media resources
4. collected evidence to support their scientific explanation

In Explain, provide students with time to formally compose their scientific explanations around the Explain Question or student-generated questions using evidence collected from Explore.

Scientific explanations are student responses, either written or orally presented, that explain scientific phenomena based upon evidence. Developing a scientific explanation requires students to analyze and interpret data to construct meaning out of the data. There are three main components to the scientific explanation: the claim, the evidence, and the reasoning.

To help students to communicate their scientific explanations, allow them to utilize the multimedia creation tools such as Board Builder and Whiteboard. Remind them that they may upload image, audio, and video files using the "attach file" option to communicate their scientific explanations.

Students may construct their scientific explanations individually or within a small group of students. Students should communicate their explanations with other classmates, and provide constructive criticism and refine their explanations prior to submission to the teacher. If explanations are used as a formative assessment, you can provide additional feedback and comments to support students as they refine their explanations.

EXPLAIN

How do we use natural resources, and why is conservation of natural resources important?

Elaborate with STEM `45–135 minutes`

*Elaborate with STEM are optional extension resources available after students have demonstrated proficiency with standards addressed previously in the concept.

NGSS Components

SEP	CCC
■ Obtaining, Evaluating, and Communicating Information ■ Asking Questions and Defining Problems ■ Planning and Carrying Out Investigation ■ Using Mathematics and Computational Thinking ■ Constructing Explanations and Designing Solutions	■ Stability and Change ■ Cause and Effect ■ Structure and Function

STEM In Action `45 minutes`

STEM in Action ties the scientific concepts to real-world applications, with many connecting to STEM careers. Technology Enhanced Items (TEIs) expect students to critically read the Core Interactive Text (CIT) and review the provided media resources.

Applying Natural Resources

- Core Interactive Text: Applying Natural Resources
- Video: Better Biofuels
- Image: Growing Energy Crops
- Video: Fuel from Algae
- Video: Solar Powered Homes
- Video: Natural Carbon Scrubbers
- Video: Drilling for Oil
- Video: Drilling for Heat

TEACHER NOTE Have students work individually to complete this short assignment. This assignment gets students thinking about renewable versus nonrenewable resources and the benefits and drawbacks of replacing traditional fuels with alternative sources of energy. It may be used as a formative or summative assessment item.

TEI Burning Biomass

STEM Project Starters

STEM Project Starters provide additional real-world contexts that require students to apply and extend their content knowledge related to the concept. STEM Project Starters can also serve as an alternative instructional hook presented at the beginning of the learning progression. The project can then be revisited throughout and at the end of the 5E learning cycle, for students to apply content knowledge.

STEM Project Starter: Reducing Your Footprint

Recommended 90 minutes

How are you reducing your use of natural resources?

TEACHER NOTE This project extends the Hands-On Activity students completed in the Explore section by having them apply reduction strategies to their daily lives. Students chart their outcome using the Graphing Calculator. Students should complete this project individually. This summative project addresses science and technology.

- Hands-on Activity: Becoming Aware of Resource Use
- Science Tool: Data/Graphing Tool

TEI Reducing Your Footprint

STEM Project Starter Make Your Home Greener

Recommended 90 minutes

How can homes be made greener?

TEACHER NOTE Ensure students have completed the Hands-On Lab Design an Off-the-Grid Home before beginning this project. Students should work individually to complete this project. This summative project addresses science, technology, and engineering.

- Hands-On Lab: Design an Off-the-Grid Home

TEI Green Home

STEM Project Starter Trickling Away

Recommended 45 minutes

Water is a natural resource that frequently can be conserved. How is water used in your community and can that use be intelligently reduced?

TEACHER NOTE Students may complete this short STEM assignment in pairs or small groups. You may wish to standardize the layout of data presentation for the class. This summative project addresses technology and math.

TEI Trickling Away

STEM Project Starter Energy Sources in Tennessee   `Recommended 90 minutes`

What energy sources provide power to Tennessee?

> **TEACHER NOTE** This STEM project starter provides students with the opportunity to research, analyze, and compare the various energy sources used in Tennessee.
>
> Students are asked to build on their understanding of renewable and nonrenewable resources from the Exploration: Natural Resources.

> **TEACHER NOTE** If students worked in small groups to complete the Exploration, then they should work with the same groups to complete this project.
>
> To expand this project, you can have student list the energy sources on the board and then divide the sources among the student groups. Students can present their findings on each source to the class before each group develops their recommendation.

- **TEI** Identify TN Energy Sources
- **TEI** Analyze Energy Sources
- **TEI** Slide Show
- **TEI** Develop Energy Recommendations

STEM Project Starter Regions with Scarce Resources   `Recommended 90 minutes`

What are solutions for regions with scarce resources?

> **TEACHER NOTE** This summative assessment provides students with the opportunity to research the resource availability of different regions on Earth. Students will then research and design a plan for potential solutions for a region with scarce resources.
>
> Students are asked to build on their understanding of energy consumption across the planet to consider global resource use and availability.

> **TEACHER NOTE** If students worked in small groups to complete the Hands-On Activity, then they should work with the same groups to complete this project.

■ Hands-On Activity: Predicting Global Energy Consumption

- **TEI** Regions with Scarce Resources
- **TEI** Discuss Proposed Solutions
- **TEI** Compare and Contrast
- **TEI** Develop a Plan

Evaluate 45–90 minutes

Explain Question:

How do we use natural resources, and why is conservation of natural resources important?

Lesson Questions (LQ):

- What are natural resources and how do we use them?

Throughout instruction and the 5E learning cycle, you will have collected formative assessment data to drive the assignment of resources and experiences to students. Evaluate is intended to include summative assessment checks for proficiency. You can use the CYE and Lesson Questions for the concept as a summative assessment in a variety of ways such as these:

- Post each Lesson Question (LQ) in various locations in the classroom, and have small groups of students generate claim statements related to the Lesson Question (LQ). Other students can add to the claim, or refute the claim, during a gallery walk where they place additional pieces of evidence on each Lesson Question (LQ) poster.
- Assign small groups of students to each Lesson Question (LQ) and have the groups generate a poster, board, graphic, or piece of text that answers the question. Use a jigsaw approach and create a second set of groups that contain members from each Lesson Question (LQ) group to share their ideas.
- Ask students to return to their initial ideas for the Explain question and add additional details and evidence.

Encourage students to review the concept review and complete the Student Self-Check practice assessment prior to assigning the Summative Teacher Concept assessment.

- Student Review and Practice Assessment
- Teacher Concept Assessments

Relationships Between Human Activity and Earth's Systems

The Five E Instructional Model

Science Techbook follows the 5E instructional model. This Model Lesson includes strategies for each of the 5Es. As you design the inquiry-based learning experience for students, be sure to collect data during instruction to drive your instructional decisions. Point-of-use teacher notes are also provided within each E-tab.

Engage   45–90 minutes

Engage Media Resources

The resources found in Engage are intended to stimulate students by exposing them to a phenomenon relevant to the content of the lesson. Engage also provides examples of relevant real-world applications that allow students to begin to make observations and relate the science content to their everyday lives. The Core Interactive Text (CIT) and media resources are carefully designed to prompt students to begin asking questions that they can investigate during the Explore phase of the lesson. They should also start collecting evidence to address the Explain question located at the bottom of the Engage page.

> **TEACHER NOTE** **Investigative Phenomenon:** One root cause of environmental problems is the rapidly growing size of the world's human population. Have students work in pairs as they examine the graph: Human Population Size. Ask them to work together to answer the following questions: (1) What factors do you think were restricting population growth in the hunter gathering and agricultural phases shown on the graph? and (2) What do you think enabled the increase in population during (a) the agricultural phase and (b) the industrial phase? Humans are mammals. For most mammals, population numbers fluctuates up and down. Sometimes populations crash. What factors could cause the human population to fluctuate? What factors could cause the population numbers to crash? Have human populations crashed in the past? If so, what was the cause?
>
> Have students work in pairs to share their ideas and debate their answers to these questions with other student paired groups. Upon completion, have students vote on the following question: Do you think that current rates of human population growth are sustainable? After voting, select individual students to justify their vote. Consider re-voting at the end of the concept. Have opinions changed? If so, why is this the case?

- Core Interactive Text: Exploring Relationships between Human Activity and Earth's Systems
- Image: Human Population Size
- Video: How Climate and Geography Affected the Evolution of Society
- Video: Agriculture and the Fertile Crescent

Explain Question

The Explain question focuses students on gathering information in the Explore section. The Explain question can be used to

- Record what students already know related to the Explain question.
- Serve as a template or model for students to generate their own scientific questions.
- Collect evidence as students work through the lesson.
- Allow students to reflect on their growth before and after the lesson.

Explain

What effect have increases in population and advancements in technology had on the way humans use Earth's natural resources?

- Image: Population of California: 1850-2000

Engage Formative Assessment

Technology Enhanced Items (TEIs) found on the Engage page enable you to collect data on students' prior knowledge and identify the common misconceptions they may possess that are related to the topic of study. These items are designed as quick checks for understanding and allow each student one attempt at each question. You can use the data collected to decide whether to assign additional resources to the class, or determine what individual or groups of students may need reinforcement or accelerated learning, prior to completing the Explore portion of the lesson.

TEACHER NOTE Student responses to this formative assessment will help evaluate prior knowledge of the concept. Students should answer individually and then discuss as a class. This activity addresses the common misconception that technological advances and human exploitation of natural resources has only had detrimental effects.

TEI Your Ideas

Before You Begin

What Do I Already Know about Relationships Between Human Activity and Earth's Systems?

TEACHER NOTE This formative assessment is designed to address the common misconception that technological advances have only been detrimental to the environment when, in fact, new technologies are being developed to protect and benefit the environment. Note that reasonable people disagree about whether some of these technologies have an overall beneficial or detrimental effect and that students may likewise disagree; they have the opportunity to explain their reasoning for their classifications.

Teachers can have students complete this item independently or with a partner. After students have completed the second item in the set, discuss answers as a whole class.

TEI Good or Bad?

TEI Explain

TEACHER NOTE This formative assessment is designed to address students' prior knowledge and understanding of renewable and nonrenewable resources. Teachers can make the connection that by using nonrenewable resources, we have an impact on the environment. Teachers may also discuss how technology can reduce the use of nonrenewable resources. Students may complete as a think-pair-share and/or discuss as a whole class. The teacher may bring up a discussion about water as a renewable resource and ask the question, *Is water (or wood) always a renewable resource?*

TEI Renewable or Nonrenewable?

TEACHER NOTE This formative activity assesses student understanding of how humans have caused problems, such as the extinction of species. It addresses the common misconception that climate and natural systems are too large for humans to affect. Teachers can have students complete as a think-pair-share or small-group activity. Teachers should encourage students to choose the best answer for each effect, so each choice is only used once. Since several of the causes may have multiple effects, correct responses may vary. After completing the activity, you should open the class for discussion to explore alternative answers and probe student's reasoning and interpretations.

TEI Human Interaction

TEACHER NOTE This formative assessment is designed to assess students' general understanding of Earth's systems and how they interact with one another. Students should complete this assignment individually.

The purpose of this activity is to stimulate student thinking toward making connections between Earth's spheres. Teachers should be aware that there could be alternative acceptable solutions to this activity, and the suggested answers provide a guide for further discussion. After students complete this activity, the teacher should open the topic to classroom discussion.

TEI Earth Sphere Interactions

- Video: Natural Resources
- Video: Interaction between Earth's Systems

Explore — 180 minutes

Lesson Questions (LQs):

1. How has an abundance of natural resources altered human activity?
2. How have Earth's systems affected human activity?
3. How have human technological designs affected the environment?
4. What relationships exist between Earth's systems and human activities?

Effective science instruction involves a student-centered rather than a teacher-centered approach. This can be accomplished either with Directed Inquiry or Guided Inquiry, depending on the needs and abilities of your class. Encourage students to select a variety of resources in their pursuit of answers as they work through Explore, with the end goal of constructing their scientific explanation in the Explain tab.

Directed Inquiry	Guided Inquiry
In Directed Inquiry, teachers provide students with a sequence of specified resources, challenging questions, and clear outcomes. Within this context students are given the opportunity to interact independently with each resource as prescribed by the teacher. Often different students' groups can be guided through several different resources at the same time. For example, one group could work on a reading passage while a second group conducts a small-group Hands-On Activity with the teacher, and a third group is independently engaged with an online interactive resource.	In Guided Inquiry, students have independence to decide the scope and sequence of their investigations. Using resources from Techbook, students determine for themselves which resources they will Explore to answer the Lesson Questions. It is important to note that each student will choose multiple resources, but no one student is expected to use all the resources available. Students also determine the order in which to explore these resources and how to record their findings.

NGSS Components

SEP	CCC
■ Analyzing and Interpreting Data ■ Obtaining, Evaluating, and Communicating Information	■ System and System Models

Lesson Question: How Has an Abundance of Natural Resources Altered Human Activity?

Recommended 25 minutes

TEACHER NOTE Connections: Crosscutting Concept: System and System Models: In this concept, students will analyze Earth's systems by defining Earth's boundaries and initial conditions, as well as its inputs and outputs. Help students think about how they interact with Earth every day. Ask them to think about the resources they have consumed thus far today. Have students choose something from their everyday life, such as the meal they had for breakfast, or the car they rode in to school. Students can model the flow of energy and matter by creating an illustrated diagram that shows how the resources involved in the object they chose move within and between Earth's systems.

- Core Interactive Text: How Has an Abundance of Natural Resources Altered Human Activity?
- Video: The Invention of the Sail and Exploration
- Video: Water and Sewer Systems
- Image: The Roman Aqueduct at Tarragona, Spain

TEACHER NOTE Misconception: Students may think that technological advances have been detrimental only to the environment. In fact, many technologies, such as solar energy, have improved the environment.

Formative Assessment:

Throughout Explore, Technology Enhanced Items (TEIs) are embedded as multi-dimensional formative checks for understanding. You can use the data they provide to

- assign additional support
- extend learning
- design additional learning tasks to clarify student misconceptions

The Explore TEIs provide students with three attempts to demonstrate their proficiency. Scaffolded feedback is provided for each attempt. If a student does not achieve proficiency by the third attempt, a media asset is provided as an additional learning opportunity.

TEACHER NOTE Practices: Science and Engineering Practice: Obtaining, Evaluating, and Communicating Information: In this item, students think about what human inventions relied on the development of other discoveries. They will read scientific literature critically to gather evidence and summarize the central ideas. Give students the opportunity to research each of the inventions mentioned in this item. Use the "That Sums It Up" strategy by assigning different groups of students a different invention mentioned in this item. Once students have completed their own research, they should present their findings with the other groups.

TEI Mining

Lesson Question: How Have Earth's Systems Affected Human Activity?

Recommended 25 minutes

- Core Interactive Text: How Have Earth's Systems Affected Human Activity?
- Sandstorms: Beneath the Layers of Ash: The People of Pompeii
- Video: Hurricane Sandy: The Impact on Society

Lesson Question: How Have Human Technological Designs Affected the Environment?

Recommended 65 minutes

- Core Interactive Text: How Have Human Technological Designs Affected the Environment?
- Video Humans' Impact on Earth
- Video Chalk Talk: Tragedy of the Commons
- Video Science in Progress: Rain Forest Removal

TEACHER NOTE There are interactions between biotic and abiotic factors in an ecosystem. Humans are not the only organisms that exploit Earth's resources. Have students consider different animals such as ants or bears and ask them how each species also uses Earth's resources.

- Image: Coal Strip Mining (3)
- Video: Mining
- Video: Human Impacts on the Oceans
- Core Interactive Text: Human Impacts on the Atmosphere
- Video: Air Pollution in Mexico City
- Video: The Ozone Layer
- Image Changes in Atmospheric Carbon Dioxide Since 1960
- Image Landfills Produce Methane

UNIT 6: Human Activities and the Biosphere

Lesson Question: What Relationships Exist Between Earth's Systems and Human Activities?

Recommended 65 minutes

- Core Interactive Text: What Relationships Exist Between Earth's Systems and Human Activities?
- Video: Hawaii and the Great Pacific Garbage Patch
- Video: Recycling Nuclear Waste
- Video: Wind Energy
- Video: Hydroelectric Power Plant of Niagara Falls
- Video: Geothermal Energy
- Video: The Electricity of Tomorrow
- Video: Coal
- Video: Biomass Energy
- Video: Core Science Concepts: Industrial Food and Livestock Production
- Video: Population Growth and the Agricultural Industry
- Video: Sustainable Agriculture
- Video: Biomagnification
- Video: Water Pollution

TEACHER NOTE **Practices: Science and Engineering Practice: Analyzing and Interpreting Data:** In this collection of items, students will interpret data to understand the impact water resources have on human activity. They will analyze data in order to make valid and reliable scientific claims. Use the "Myth Bustin'!" strategy to help students interpret data. Provide a series of statements about each graph and have students determine whether the statements are supported by the graph. The "Myth Bustin'!" strategy is found on the Professional Learning tab. Click on Strategies & Resources, then Spotlight On Strategies (SOS). "Myth Bustin'!" is found underneath "Summarizing."

TEI Water Use

TEI Climate Change

Explore More Resources

Resources in Explore More Resources support differentiation within your classroom by

- providing additional visualization of content
- affording extension of content to those students ready for acceleration
- offering Lexile reading levels for reading passages

Online explorations and hands-on experiences are provided so that students can conduct virtual investigations, collect and design investigations, and collect and analyze data; these skills are essential to developing scientific understanding.

Explain `45–90 minutes`

In Explore, students
1. uncovered scientific understandings
2. conducted investigations
3. analyzed data, text, and other media resources
4. collected evidence to support their scientific explanation

In Explain, provide students with time to formally compose their scientific explanations around the Explain or student-generated questions using evidence collected from Explore.

Scientific explanations are student responses, either written or orally presented, that explain scientific phenomena based upon evidence. Developing a scientific explanation requires students to analyze and interpret data to construct meaning out of the data. There are three main components to the scientific explanation: the claim, the evidence, and the reasoning.

To help students to communicate their scientific explanations, allow them to utilize the multimedia creation tools such as Board Builder and Whiteboard. Remind them that they may upload image, audio, and video files using the "attach file" option to communicate their scientific explanations.

Students may construct their scientific explanations individually or within a small group of students. Students should communicate their explanations with other classmates, and provide constructive criticism and refine their explanations prior to submission to the teacher. If explanations are used as a formative assessment, you can provide additional feedback and comments to support students as they refine their explanations.

EXPLAIN

What effects have increases in population and advancements in technology had on the way humans use Earth's natural resources?

Elaborate with STEM   45–135 minutes

*Elaborate with STEM are optional extension resources available after students have demonstrated proficiency with standards addressed previously in the concept.

NGSS Components

SEP	CCC
■ Obtaining, Evaluating, and Communicating Information ■ Asking Questions and Defining Problems ■ Planning and Carrying Out Investigation ■ Using Mathematics and Computational Thinking ■ Constructing Explanations and Designing Solutions	■ Stability and Change ■ Cause and Effect ■ Structure and Function

STEM In Action   45 minutes

STEM in Action ties the scientific concepts to real-world applications, with many connecting to STEM careers. Technology Enhanced Items (TEIs) expect students to critically read the Core Interactive Text (CIT) and review the provided media resources.

Applying Relationships between Human Activity and Earth's Systems

- Core Interactive Text: Applying Relationships between Human Activity and Earth's Systems
- Video: Searching for Oil beneath the Ocean Floor
- Video: Balancing Conservation and Industrial Growth
- Video: The Grain of the Future
- Video: Sustainable Fishing Methods
- Video: Preventing Landslides

TEACHER NOTE This summative activity assesses students' understanding of increases in population growth rates and how advances in technology have played a role in those rates. Teachers should also discuss with students the effects of increased populations on Earth's systems.

TEI Human Population

- Growth Rate

STEM Project Starters

STEM Project Starters provide additional real-world contexts that require students to apply and extend their content knowledge related to the concept. STEM Project Starters can also serve as an alternative instructional hook presented at the beginning of the learning progression. The project can then be revisited throughout and at the end of the 5E learning cycle, for students to apply content knowledge.

STEM Project Starter: Helping Farmers

Recommended 60 minutes

Why is preventing erosion important?

> **TEACHER NOTE** This summative assessment is designed to assess students' understanding of some of the processes designed to decrease the impact of humans on the environment, particularly related to soil conservation. Students need to conduct research about soil conservation methods to successfully complete this activity. Students should complete the activity individually. Teachers can expand the activity by having groups of students each research a different conservation method and give a presentation on the method to the class.

TEI Helping Farmers

STEM Project Starter Technology to Conserve Energy

Recommended 90 minutes

How does technology work to conserve energy resources?

> **TEACHER NOTE** This summative assessment is designed to assess students' understanding of types of technology used to decrease the effects of human interaction on the environment. Teachers may have students complete this activity in small groups with each group taking on a different type of technology. Students may present their diagrams and findings using Board Builder.

TEI Research Report

TEI Diagram

- Activity: Engineering Design Sheet

Evaluate `45–90 minutes`

Explain question:
What effects have increases in population and advancements in technology had on the way humans use Earth's natural resources?

Lesson Questions (LQ):
- How has an abundance of natural resources altered human activity?
- How have Earth's systems affected human activity?
- How have human technological designs affected the environment?
- What relationships exist between Earth's systems and human activities?

Throughout instruction and the 5E learning cycle, you will have collected formative assessment data to drive the assignment of resources and experiences to students. Evaluate is intended to include summative assessment checks for proficiency. You can use the Explain and Lesson Questions for the concept as a summative assessment in a variety of ways such as these:

- Post each Lesson Question (LQ) in various locations in the classroom, and have small groups of students generate claim statements related to the Lesson Question (LQ). Other students can add to the claim, or refute the claim, during a gallery walk where they place additional pieces of evidence on each Lesson Question (LQ) poster.
- Assign small groups of students to each Lesson Question (LQ) and have the groups generate a poster, board, graphic, or piece of text that answers the question. Use a jigsaw approach and create a second set of groups that contain members from each Lesson Question (LQ) group to share their ideas.
- Ask students to return to their initial ideas for the Explain question and add additional details and evidence.

Encourage students to review the concept review and complete the Student Self-Check practice assessment prior to assigning the Summative Teacher Concept assessment.

- Student Review and Practice Assessment
- Teacher Concept Assessments

Understanding Climate and Climate Change

The Five E Instructional Model

Science Techbook follows the 5E instructional model. This Model Lesson includes strategies for each of the 5Es. As you design the inquiry-based learning experience for students, be sure to collect data during instruction to drive your instructional decisions. Point-of-use teacher notes are also provided within each E-tab.

Engage   45–90 minutes

Engage Media Resources

The resources found in Engage are intended to stimulate students by exposing them to a phenomenon relevant to the content of the lesson. Engage also provides examples of relevant real-world applications that allow students to begin to make observations and relate the science content to their everyday lives. The Core Interactive Text (CIT) and media resources are carefully designed to prompt students to begin asking questions that they can investigate during the Explore phase of the lesson. They should also start collecting evidence to address the Explain question located at the bottom of the Engage page.

TEACHER NOTE **Investigative Phenomenon:** To stimulate student thinking about climate and changes in climate, focus on how climate impacts their lives. For example: How does the existence and nature of seasons affect the clothes people buy or the design of their homes? How do human activities, such as sports and other form of recreation, vary with the seasons? How might these be different if they lived in another climate where the seasons were different? Use the video Climate and Its Impact on Human Activities to get students thinking further about this. What would happen if the climate where they live changed from what it is now? For example, if it got much hotter, drier, or much wetter with more storms or torrential rain, how would such changes impact their lives and lives of other people?

- Core Interactive Text: Thinking About Climate and Climate Change
- Video: Climate and its Impacts on Human Activities
- Image: U.S. Climate Regions
- Video: Global Warming and Weather
- Video: Climate

UNIT 6: Human Activities and the Biosphere

Explain Question

The Explain question focuses students on gathering information in the Explore section. The Explain question can be used to

- Record what students already know related to the Explain question.
- Serve as a template or model for students to generate their own scientific questions.
- Collect evidence as students work through the lesson.
- Allow students to reflect on their growth before and after the lesson.

Explain

How do human activities change the world's climate?

- Image: A Pulp Mill in Reversing Falls of the St. John River in St. John, New Brunswick, Canada

Engage Formative Assessment

Technology Enhanced Items (TEIs) found on the Engage page enable you to collect data on students' prior knowledge and identify the common misconceptions they may possess that are related to the topic of study. These items are designed as quick checks for understanding and allow each student one attempt at each question. You can use the data collected to decide whether to assign additional resources to the class, or determine what individual or groups of students may need reinforcement or accelerated learning, prior to completing the Explore portion of the lesson.

TEACHER NOTE Use this formative assessment to evaluate students' basic knowledge of the nature of the greenhouse effect.

Suggested use of this response includes students working in think-pair-share groups to discuss their thoughts about the greenhouse effect.

TEI Your Ideas

Before You Begin

What do I already know climate and climate change?

> **TEACHER NOTE** This formative activity is intended to address student misconceptions about human activities that impact global warming.
>
> Students may believe that human activity cannot change climate because the atmosphere and Earth itself are so big. In fact, human activity has changed many of Earth's systems. These changes have accelerated since the Industrial Revolution. Currently, there are around 7 billion people on Earth. The emissions of greenhouse gases from the activities of so many people have resulted in a measurable increase in concentrations of carbon dioxide and other greenhouse gases in the atmosphere. The activity can be performed in think-pair-share groups.

TEI Classifying Human Activities

- Video: Transforming Energy
- Video: What is Matter?
- Video: What is in the Air?
- Video: Ocean Currents and Climate
- Video: What Causes Weather?

Explore `315 minutes`

Lesson Questions (LQs):
1. What factors determine climate?
2. What processes are involved in climate change?

Effective science instruction involves a student-centered rather than a teacher-centered approach. This can be accomplished either with Directed Inquiry or Guided Inquiry, depending on the needs and abilities of your class. Encourage students to select a variety of resources in their pursuit of answers as they work through Explore, with the end goal of constructing their scientific explanation in the Explain tab.

Directed Inquiry	Guided Inquiry
In Directed Inquiry, teachers provide students with a sequence of specified resources, challenging questions, and clear outcomes. Within this context students are given the opportunity to interact independently with each resource as prescribed by the teacher. Often different students groups can be guided through several different resources at the same time. For example, one group could work on a reading passage while a second group conducts a small-group Hands-On Activity with the teacher, and a third group is independently engaged with an online interactive resource.	In Guided Inquiry, students have independence to decide the scope and sequence of their investigations. Using resources from Tech book, students determine for themselves which resources they will Explore to answer the Lesson Questions. It is important to note that each student will choose multiple resources, but no one student is expected to use all the resources available. Students also determine the order in which to explore these resources and how to record their findings.

NGSS Components

SEP	CCC
■ Analyzing and Interpreting Data ■ Developing and Using Models ■ Scientific Investigations Use a Variety of Methods ■ Scientific Knowledge is Based on Empirical Evidence ■ Engaging in Argument from Evidence	■ Cause and Effect ■ Influence of Engineering, Technology, and Science on Society and the Natural World ■ Energy and Matter ■ Stability and Change

Lesson Question: What factors determine climate?

Recommended 90 minutes

TEACHER NOTE Science and Engineering Practice: Engaging in Argument from Evidence: Throughout this concept, students will learn about how climate relates to and is different from weather, as well as the factors that determine climate. Students will also learn about both natural and human processes that lead to climate change and will discuss changes in climate that have happened in the past and that are happening in the present. Tell students that they will be engaging in the argument over the human impact of climate change by using appropriate and sufficient evidence to choose a position on how and to what extent human activity impacts climate change.

Suggest to students that they create a table with human and natural causes of climate change and that throughout the lesson they should write evidence in each column. Then, at the end of the lesson, they should use this evidence to present a case about the human impact on climate change. As students read and comprehend complex texts, view the videos, and complete the interactives, labs, and other Hands-On Activities, have them summarize and obtain scientific and technical information. Students will use this evidence to support their initial ideas on how to answer the Explain Question or their own question they generated during Engage. Have students record their evidence using My Notebook

- Core Interactive Text: What Factors Determine Climate?
- Video: Weather and Climate
- Image: Average Temperatures in Memphis Tennessee
- Image: Average Precipitation in Memphis Tennessee
- Video: Climate
- Exploration: Understanding Climate

Formative Assessment:

Throughout Explore, Technology Enhanced Items (TEIs) are embedded as multi-dimensional formative checks for understanding. You can use the data they provide to

- assign additional support
- extend learning
- design additional learning tasks to clarify student misconceptions

The Explore TEIs provide students with three attempts to demonstrate their proficiency. Scaffolded feedback is provided for each attempt. If a student does not achieve proficiency by the third attempt, a media asset is provided as an additional learning opportunity.

TEACHER NOTE Science and Engineering Practice: Analyzing and Interpreting Data: In this item, students will use the graphs shown in the Media Gallery to evaluate information about how the average temperature and monthly precipitation in Memphis changes over the course of a year and then synthesize that information by completing a series of dropdowns. Tell students to examine the graphs. Then, ask them what conclusions they can draw about temperature and precipitation patterns in Memphis. Ask them how stable they think these patterns are. Then have them write down some ideas to express what factors might change that stability. Have students create a projected timeline in which they graph what they think Memphis's climate will look like over the next ten years, giving reasons for any changes they predict.

TEI A Year in Memphis

TEACHER NOTE Connections: Crosscutting Concept: Energy and Matter: Have students focus on the image Earth's Energy Budget as they begin to build a model as to how the flow of energy into, between, and out of Earth's systems determine climatic conditions and how changes in Earth's energy balance could result in changes in climate both on a regional and global scale.

- Image: Earth's Energy Budget
- Video: Solar Thermal Energy in Earth
- Image: Rain Shadow
- Video: The Ocean and Climate
- Image: Wind Circulation

TEACHER NOTE Connections: Crosscutting Concept: Cause and Effect: In this item, students will suggest cause-and-effect relationships between the various factors that affect a region's climate. Point out that all of these factors are interrelated and ask students to draw a chart showing how each factor connects with the others. Ask them to explain and predict what would happen to a region's climate if one of the smaller-scale mechanisms within the larger system suddenly changed (if a volcano erupted). What would this change about each of the factors they have studied, and how would each change affect the others? Extend the conversation from local to global climate causes and effects.

TEI A Little Latitude
TEI Cold Factors

- Video: Causes of the Seasons
- Video: Geography and Climate
- Hands-On Lab: Using Climate Models to Forecast Impacts of Climate Change
- Image: Global Climate Zone Map
- Video: Climate and Climate Change

6.3 Understanding Climate and Climate Change

Lesson Question: What Processes Are Involved in Climate Change?

Recommended 45 minutes

TEACHER NOTE Misconception: Students may believe that all climate change is a result of human activities. In fact, local and global climate change has occurred throughout Earth's history as a result of natural processes like plate tectonics and changes in Earth's tilt and orbit. One reason that the issue of climate change is so controversial is because it can be difficult to separate the effects of natural processes on climate from the effects of human activities on climate.

- Core Interactive Text: What Processes Are Involved in Climate Change?
- Image: Rings in a Tree
- Video: Reading Ocean Sediment Cores
- Video: Ice Cores
- Video: Algae and Climate Change

TEACHER NOTE You might want to break up the following list of climate change factors below, following each one with some discussion. Or, assign pairs of students to each factor and have them read their paragraph, then explain the factor to the class using a jigsaw method. Have pairs explain what they think it means and how it impacts climate change.

- Video: Peru: El Niño Devastation
- Video: Sulphur Dioxide

TEACHER NOTE Misconception: This item addresses the misconception that cold or unusually cool weather is proof that global warming is not really happening. Students with this misconception confuse weather with climate. In fact, global warming refers to overall changes in Earth's climate, which describes average atmospheric conditions over long periods of time. Evidence demonstrates that the average temperature of the atmosphere has been increasing over recent decades. Locally variable weather conditions are distinct from the overall trend toward warming. In addition, as average global temperatures increase, local or regional climates can become cooler. A suggested use of the activity involves a whole-class or small-group discussion, in which each statement is discussed by students to help formulate their answers.

- Reading Passage: Little Ice Age

TEI Cold Weather

TEACHER NOTE Misconception: Students may believe that cold or unusually cool weather is proof that global warming is not really happening. Students with this misconception confuse weather with climate. In fact, global warming refers to overall changes in Earth's climate, which describe average atmospheric conditions over long periods of time. Evidence demonstrates that the average temperature of the atmosphere has been increasing over recent decades. Locally variable weather conditions are distinct from the overall trend toward warming. In addition, as average global temperatures increase, local or regional climates can become cooler. Scientists predict, for example, that global warming will affect warm ocean currents, changing their routes or cutting them off completely. As a result, places that are now warm because of these currents may become cooler.

- Image: Carbon Dioxide Concentration in the Atmosphere
- Image: Anthropogenic Greenhouse Gas Emissions

TEACHER NOTE Misconception: Students may believe that human activity cannot change climate because the atmosphere and Earth itself are so big. In fact, human activity has changed many of Earth's systems. These changes have accelerated since the Industrial Revolution. Currently, there are around seven billion people on Earth. The emissions of greenhouse gases from the activities of so many people have resulted in a measurable increase in concentrations of carbon dioxide and other greenhouse gases in the atmosphere.

- Video: Impacts of Global Warming
- Video: Tennessee Wildfire
- Video: The Greenhouse Effect
- Video: Global Climate and Changes in Ecosystems
- Image: Glacier Calving
- Video: Shrinking Glaciers
- Reading Passage: The Calving of A68
- Video: Greenland's Retreating Glaciers
- Video: Snowball Earth
- Video: Introduction: Snowball Earth
- Reading Passage: Feedback Effects on Climate

TEACHER NOTE Science and Engineering Practice: Developing and Using Models: In this item, students will develop a complex model that allows for the testing of the processes that influence climate change. Tell students that they will be putting into practice what they have learned about the factors that influence climate change. Ask them to draw a diagram showing how each of the factors they have studied affects the overall climate of Earth. Remind them of their table showing human and natural factors they have been tracking and writing about throughout this concept. Have them present their findings and decisions to the class about the role humans play in climate change.

TEI Factors behind Climate Change

- Image: Climate Model

Explain `45–90 minutes`

In Explore, students
1. uncovered scientific understandings
2. conducted investigations
3. analyzed data, text, and other media resources
4. collected evidence to support their scientific explanation

In Explain, provide students with time to formally compose their scientific explanations around the Explain or student-generated questions using evidence collected from Explore.

Scientific explanations are student responses, either written or orally presented, that explain scientific phenomena based upon evidence. Developing a scientific explanation requires students to analyze and interpret data to construct meaning out of the data. There are three main components to the scientific explanation: the claim, the evidence, and the reasoning.

To help students to communicate their scientific explanations, allow them to utilize the multimedia creation tools such as Studio and Whiteboard. Remind them that they may upload image, audio, and video files using the "attach file" option to communicate their scientific explanations.

Students may construct their scientific explanations individually or within a small group of students. Students should communicate their explanations with other classmates, and provide constructive criticism and refine their explanations prior to submission to the teacher. If explanations are used as a formative assessment, you can provide additional feedback and comments to support students as they refine their explanations.

EXPLAIN
How do human activities change the world's climate?

Elaborate with STEM `45–135 minutes`

*Elaborate with STEM are optional extension resources available after students have demonstrated proficiency with standards addressed previously in the concept.

NGSS Components

SEP	CCC
■ Asking Questions and Defining Problems ■ Developing and Using Models ■ Analyzing and Interpreting Data ■ Using Mathematics and Computational Thinking	■ Cause and Effect ■ Stability and Change ■ Systems and System Models

STEM In Action `45 minutes`

STEM in Action ties the scientific concepts to real-world applications, with many connecting to STEM careers. Technology Enhanced Items (TEIs) expect students to critically read the Core Interactive Text (CIT) and review the provided media resources.

Applying Understanding Climate and Climate Change

- Core Interactive Text: Applying Understanding Climate and Climate Change
- Image: The Impact of Glaciers
- Video: The Threat of Global Warming
- Image: University of California Marine Biologists Gathering Krill Samples, Monterey Bay, California
- Video: Conducting Ice Research

TEACHER NOTE This summative response is intended to stimulate student learning about technologies for studying climate. Teachers can implement the activity in think-pair-shares or small groups.

- **TEI** Reading Passage: Global Climate Models
- **TEI** Climate Models
- **TEI** Tennessee Climate Model

STEM Project Starters

STEM Project Starters provide additional real-world contexts that require students to apply and extend their content knowledge related to the concept. STEM Project Starters can also serve as an alternative instructional hook presented at the beginning of the learning progression. The project can then be revisited throughout and at the end of the 5E learning cycle, for students to apply content knowledge.

STEM Project Starter: Modeling the Effects of Climate Change `Recommended 45 minutes`

What types of ecosystems exist across the Earth?

> **TEACHER NOTE** Teachers can use this summative activity to assess students' ability to model the effects of climate change. Teachers can implement the activity in think-pair-shares or small groups.
>
> Students should watch the video segment to prompt their thoughts about ecosystems.

- Video: Marine Ecosystem Diversity

TEI Climate Change Models

STEM Project Starter Tracking Greenhouse Gas Emissions `Recommended 45 minutes`

What can you do to reduce your carbon footprint?

> **TEACHER NOTE** Teachers can use this summative activity to assess student knowledge of carbon footprints. Teachers can implement the activity in think-pair-shares or small groups, as well as independently. Teachers can help students understand the context of the video segment in terms of how transportation is affecting the carbon footprint of cities across the world.

- Hands-On Activity: Track Your Greenhouse Emissions
- Video: Future Transportation Problem

TEI Carbon Footprint

Evaluate `45–90 minutes`

Explain Question:

How do human activities change the world's climate?

Lesson Questions (LQ):

- What factors determine climate?
- What processes are involved in climate change?

Throughout instruction and the 5E learning cycle, you will have collected formative assessment data to drive the assignment of resources and experiences to students. Evaluate is intended to include summative assessment checks for proficiency. You can use the Explain and Lesson Questions for the concept as a summative assessment in a variety of ways such as these:

- Post each Lesson Question (LQ) in various locations in the classroom, and have small groups of students generate claim statements related to the Lesson Question (LQ). Other students can add to the claim, or refute the claim, during a gallery walk where they place additional pieces of evidence on each Lesson Question (LQ) poster.
- Assign small groups of students to each Lesson Question (LQ) and have the groups generate a poster, board, graphic, or piece of text that answers the question. Use a jigsaw approach and create a second set of groups that contain members from each Lesson Question (LQ) group to share their ideas.
- Ask students to return to their initial ideas for the Explain question and add additional details and evidence.

Encourage students to review the concept review and complete the Student Self-Check practice assessment prior to assigning the Summative Teacher Concept assessment.

- Student Review and Practice Assessment
- Teacher Concept Assessments

Impacts on Biodiversity

The Five E Instructional Model

Science Techbook follows the 5E instructional model. This Model Lesson includes strategies for each of the 5Es. As you design the inquiry-based learning experience for students, be sure to collect data during instruction to drive your instructional decisions. Point-of-use teacher notes are also provided within each E-tab.

Engage 45–90 minutes

Engage Media Resources

The resources found in Engage are intended to stimulate students by exposing them to a phenomenon relevant to the content of the lesson. Engage also provides examples of relevant real-world applications that allow students to begin to make observations and relate the science content to their everyday lives. The Core Interactive Text (CIT) and media resources are carefully designed to prompt students to begin asking questions that they can investigate during the Explore phase of the lesson. They should also start collecting evidence to address the Explain question located at the bottom of the Engage page.

> **TEACHER NOTE** **Investigative Phenomenon:** Before students start Engage, ask them what the term *extinction* means. What causes extinction? What is the frequency of species extinction (one a month, a year, a decade or a century)? They may find it difficult to identify examples of organisms that have recently become extinct. In Engage, they encounter the phenomena of recent examples of extinction or species close to extinction. In particular, they study the extinction of amphibians in some detail. Return to these questions at the end of Engage.

- Core Interactive Text: The Last Male
- Video: The Last Male Northern White Rhino Dies
- Video: Adaptable Amphibians
- Video: Mountain Yellow Legged Frog
- Video: Tropical Frogs Ex Situ
- Video: Threats to Biodiversityv

Explain Question

The Explain question focuses students on gathering information in the Explore section. The Explain question can be used to

- Record what students already know related to the Explain question.
- Serve as a template or model for students to generate their own scientific questions.
- Collect evidence as students work through the lesson.
- Allow students to reflect on their growth before and after the lesson.

Explain
What can humans do to reduce the extinction rate?

■ Image: Park Ranger Holding Alligator Skin

Engage Formative Assessment
Technology Enhanced Items (TEIs) found on the Engage page enable you to collect data on students' prior knowledge and identify the common misconceptions they may possess that are related to the topic of study. These items are designed as quick checks for understanding and allow each student one attempt at each question. You can use the data collected to decide whether to assign additional resources to the class, or determine what individual or groups of students may need reinforcement or accelerated learning, prior to completing the Explore portion of the lesson.

TEACHER NOTE Use student responses as a formative assessment to evaluate students' prior knowledge of biodiversity and the impact of humans. Suggested use of this item includes students working in think-pair-share groups to discuss their thoughts about biodiversity, causes of extinction, and preventing extinctions.

TEI My Ideas About Extinction Rate

Before You Begin
What Do I already know about impacts on biodiversity?

TEACHER NOTE This item assesses student understanding of the term *biodiversity*. Upon completion, encourage students to share their answers and come to consensus regarding a definition.

TEI Biodiversity

TEACHER NOTE This item assesses student understanding of the frequency of mass extinction events, their cause—rapidly changing environmental conditions—and their impact on the number of species or groups of organisms on Earth. Upon completion have students look at where the graph ends and ask them what they think the graph would look like if it was plotted up to the present day.

TEI Extinction Rates

TEACHER NOTE This item assesses student understanding of how disruption of food webs, by the removal or decline of some species in an ecosystem, has a direct and indirect impact on other species, resulting in perhaps unforeseen changes in species composition and ecosystem function.

TEI Coral Reef Diversity

- Video: Mass Extinctions
- Video: Loss of Bees
- Video: Defining Biodiversity

Explore   `270 minutes`

Lesson Questions (LQs)

1. Why does biodiversity matter?
2. What is the relationship between human population size and rates of extinction?
3. What are the main causes of biodiversity loss?
4. What steps can humans take to prevent the loss of biodiversity?

Effective science instruction involves a student-centered rather than a teacher-centered approach. This can be accomplished either with Directed Inquiry or Guided Inquiry, depending on the needs and abilities of your class. Encourage students to select a variety of resources in their pursuit of answers as they work through Explore, with the end goal of constructing their scientific explanation in the Explain tab.

Directed Inquiry	Guided Inquiry
In Directed Inquiry, teachers provide students with a sequence of specified resources, challenging questions, and clear outcomes. Within this context students are given the opportunity to interact independently with each resource as prescribed by the teacher. Often different student groups can be guided through several different resources at the same time. For example, one group could work on a reading passage while a second group conducts a small-group Hands-On Activity with the teacher, and a third group is independently engaged with an online interactive resource.	In Guided Inquiry, students have independence to decide the scope and sequence of their investigations. Using resources from Tech book, students determine for themselves which resources they will Explore to answer the Lesson Questions. It is important to note that each student will choose multiple resources, but no one student is expected to use all the resources available. Students also determine the order in which to explore these resources and how to record their findings.

NGSS Components

SEP	CCC
■ Analyzing and Interpreting Data ■ Asking Questions and Defining Problems ■ Obtaining, Evaluating, and Communicating information ■ Constructing Explanations and Designing Solutions ■ Using Mathematics and Computational Thinking ■ Engaging in Argument from Evidence	■ Cause and Effect ■ Systems and System Models ■ Stability and Change

Lesson Question: Why Does Biodiversity Matter?

Recommended 45 minutes

TEACHER NOTE The focus of this concept is the increasing number of extinctions occurring because of human activities. Most students will be unaware of the rate at which species extinction is increasing and the impact this will have on the biological systems that support life on Earth. Initially students are introduced to the extent of this issue and then progress to its causes and possible solutions. As students encounter examples of extinction have them think about how the removal or decline of even one species can create a ripple effect within and beyond the ecosystem in which it lives. Encourage them to scale up these extinctions to imagine what life would be like without these species and the ecosystems they provide. Some students may be familiar with science fiction writers and film-makers who have tried to envisage such futures, so you may be able to use some of these fictional scenarios to initiate discussion.

As students read and comprehend complex texts, view the videos, and complete the interactives, labs, and other Hands-On Activities, have them summarize and obtain scientific and technical information. Students will use this evidence to support their initial ideas on how to answer the Explain Question or their own question they generated during Engage. Have students record their evidence using "My Notebook."

- Core Interactive Text: Why Does Biodiversity Matter?
- Video: Biodiversity

TEACHER NOTE Connections: Crosscutting Concepts: Stability and Change: Use the media gallery below to stimulate student interest and discussion about the importance of biodiversity in maintaining the balance of Earth's systems and supporting human life on the planet. Get the students to think about examples, like the monkeys mentioned in the video caption below, in which a species plays a key but often unrecognized role.

- Video: Mangrove Swamps
- Video: Ocean Microbes
- Video: Monkeys and World Climate
- Video: Cave Painting and the Natural World

Formative Assessment:

Throughout Explore, Technology Enhanced Items (TEIs) are embedded as multi-dimensional formative checks for understanding. You can use the data they provide to

- assign additional support
- extend learning
- design additional learning tasks to clarify student misconceptions

The Explore TEIs provide students with three attempts to demonstrate their proficiency. Scaffolded feedback is provided for each attempt. If a student does not achieve proficiency by the third attempt, a media asset is provided as an additional learning opportunity.

TEACHER NOTE This item students apply their knowledge of groups of organisms, their function within ecosystems, and the role of biodiversity in supporting natural systems and humans that rely on them. Before students start, suggest that they begin by thinking about how each type of organism obtains the energy and nutrients it needs to live.

TEI Biodiversity Importance

Lesson Question: What is the Relationship between Human Population Size and Rates of Extinction?

Recommended 90 minutes

TEACHER NOTE Misconception: Students and many individuals may have the misconception that technology will enable the expansion of human population on Earth to continue indefinitely. Although some earlier projections of the limits to population growth were inaccurate, the destructive consequences of such growth on natural systems and their biodiversity is well documented. Eventually, without stabilization at a sustainable level, human population will reach the carrying capacity of Earth, with the inevitability of ensuing ecological collapse.

- Core Interactive Text: What is the Relationship between Human Population Size and Rates of Extinction?
- Video: Early Hunters and America's Mega Fauna
- Image: Human Population Growth over the Last 15,000 Years
- Video: The Origins of Modern Crops

TEACHER NOTE Connections: Crosscutting Concepts: Cause and Effect: Use the resources below to help students make connections between the products they consume and impacts on global biodiversity. Once students have viewed the media, have them investigate another product they commonly use and determine the direct and indirect environmental impacts it can have. Can they suggest a more biodiversity friendly product they could use in its place?

- Video: Toothpaste Endangers Orangutans
- Video: Eggs and the Ocean

6.4 Impacts on Biodiversity

TEACHER NOTE **Practices: Science and Engineering Practices: Analyzing and Interpreting Data:** In this item students analyze and interpret graphical data of human population and extinction. Once students have completed the item, have them use extrapolation to determine the size of earth's population by the year 2100. Use this as the basis for a discussion about whether a human population of this size could be sustained by Earth's ecosystems.

TEI Extinction Rates and Human Population

Lesson Question: What Are the Main Causes of Biodiversity Loss?

Recommended 90 minutes

- Core Interactive Text: What Are the Main Causes of Biodiversity Loss?

TEACHER NOTE **Practices: Science and Engineering Practices: Obtaining, Evaluating, and Communicating information:** Below are four case studies about biodiversity losses and how they can occur. All of them look at more than the loss of more than one species, not always to extinction. Each case study provides an example of the type of cascade effect that occurs in ecosystems when they are disrupted in some way, and all but one example involves multiple causes. All are related to human activities (the most common one being the introduction of invasive organisms). Depending on the time available, have students work in groups and analyze a couple or all of the case studies provided. Next have them identify, research, and evaluate another example of biodiversity loss and the ecosystems, organisms and human activities involved. Have students communicate their findings to other groups using an appropriate medium.

- Image: The Brown Tree Snake
- Image: One of the Rarest Birds in the United States
- Video: New Bird Species Evolve on the Hawaiian Islands
- Video: Tool-Using Crows
- Image: Early Loggers and Very Strong Horses
- Video: Tree Diseases
- Image: Dying Trees
- Video: Emerald Ash Borer
- Image: Tree Damaged by Acid Rain
- Video: Climate Change and Invasive Species
- Image: Tiger Shark
- Video: Stopping Shark Fin Soup
- Image: Simplified Ocean Food Chain

TEACHER NOTE Use the media in the gallery below to stimulate student thinking about other cases of biodiversity loss.

- Video: Zebra Mussels: An Invasive Species
- Video: Changes in the Arctic
- Video: One of the Most Biodiverse Areas of the World
- Hands-On Lab: How Does Trampling Impact Plant Species Composition, Diversity, and Percentage Growth?

Lesson Question: What Steps Can Humans Take to Prevent the Loss of Biodiversity?

`Recommended 45 minutes`

TEACHER NOTE **Practices: Science and Engineering Practices: Asking Questions and Defining Problems:** Prior to commencing this lesson question, consider organizing a field trip to a local nature reserve. On the visit, have students ask questions and collect information from reserve staff about the reasons for establishing the reserve, reserve management, the biodiversity of the reserve compared with the surrounding areas and specific threats (past and present), as well as ways in which the reserve objectives could be better met (for example, increasing its area, removal of invasive species, scientific research, breeding programs, etc.).

- Core Interactive Text: What Steps Can Humans Take to Prevent the Loss of Biodiversity?
- Video: Biodiversity Hot Spots
- Video: Forested Wetlands and Ecosystem Services
- Video: Sustainability and Natural Resources
- Video: Growing Food in Cities
- Video: Energy Consumption around the World
- Video: Invasion of the Earthworms

`TEI` **Modifying Biodiversity**

`TEI` **Conserving Biodiversity**

Explore More Resources

Resources in Explore More Resources support differentiation within your classroom by

- providing additional visualization of content
- affording extension of content to those students ready for acceleration
- offering Lexile reading levels for reading passages

Online explorations and hands-on experiences are provided so that students can conduct virtual investigations, collect and design investigations, and collect and analyze data; these skills are essential to developing scientific understanding.

Explain `45–90 minutes`

In Explore, students
1. uncovered scientific understandings
2. conducted investigations
3. analyzed data, text, and other media resources
4. collected evidence to support their scientific explanation

In Explain, provide students with time to formally compose their scientific explanations around the Explain or student-generated questions using evidence collected from Explore.

Scientific explanations are student responses, either written or orally presented, that explain scientific phenomena based upon evidence. Developing a scientific explanation requires students to analyze and interpret data to construct meaning out of the data. There are three main components to the scientific explanation: the claim, the evidence, and the reasoning.

To help students to communicate their scientific explanations, allow them to utilize the multimedia creation tools such as Studio and Whiteboard. Remind them that they may upload image, audio, and video files using the "attach file" option to communicate their scientific explanations.

Students may construct their scientific explanations individually or within a small group of students. Students should communicate their explanations with other classmates, and provide constructive criticism and refine their explanations prior to submission to the teacher. If explanations are used as a formative assessment, you can provide additional feedback and comments to support students as they refine their explanations.

EXPLAIN

What can humans do to reduce the extinction rate?

Elaborate with STEM   `45–135 minutes`

*Elaborate with STEM are optional extension resources available after students have demonstrated proficiency with standards addressed previously in the concept.

NGSS Components

SEP	CCC
■ Analyzing and Interpreting Data ■ Obtaining, Evaluating, and Communicating Information ■ Planning and Carrying Out Investigations ■ Using Mathematics and Computational Thinking	■ Cause and Effect ■ Stability and Change

STEM In Action   `45 minutes`

STEM in Action ties the scientific concepts to real-world applications, with many connecting to STEM careers. Technology Enhanced Items (TEIs) expect students to critically read the Core Interactive Text (CIT) and review the provided media resources.

Applying Impacts on Biodiversity

- Core Interactive Text: Applying Impacts on Biodiversity
- Video: Invasive Species
- Video: A Threat to Tasmanian Devils
- Video: Forest Fragmentation
- Video: Mass Extinctions
- Video: Impact of Human Activities
- Video: The Impact of Gold Mining

TEACHER NOTE Use the "Threatened Species" item as a formative assessment to ensure that students understand that scientists can quantify the extinction risk of species and assign a species to a specific category of extinction risk. Use the "Summarizing Human Impacts" item as a summative assessment to determine student understanding of the connection between the effects of human activities on species' populations and extinction risk, and of the potential for solutions.

TEI Threatened Species
TEI Summarizing Human Species

STEM Project Starters

STEM Project Starters provide additional real-world contexts that require students to apply and extend their content knowledge related to the concept. STEM Project Starters can also serve as an alternative instructional hook presented at the beginning of the learning progression. The project can then be revisited throughout and at the end of the 5E learning cycle, for students to apply content knowledge.

STEM Project Starter Species Richness   Recommended 135 minutes

How do ecologists calculate species diversity for a specific area or ecosystem?

- Image: Wild Flowers

TEACHER NOTE Have students complete the Hands-On Lab Classifying Species in the Field before attempting this project.

In this summative assessment, students will demonstrate their ability to calculate species diversity using a well-accepted diversity index.

Students may need help calculating H. Remind them that they need to do the proportion calculation for each species and then sum across all species to get H.

Class Use: Students may work in pairs or small groups to collect data and calculate H scores. All data can be shared with the class, and a discussion could follow relating to the relative diversity or paucity of species in the area.

- Core Interactive Text: Species Richness
- Hands-On Lab: Classifying Species in the Field

TEI Calculating Diversity

STEM Project Starter A Floating Dump   Recommended 90 minutes

What efforts can be made to reduce the impact of trash in the oceans?

- Image: Bottles Floating in Drainage

TEACHER NOTE This STEM project is a formative assessment related to the study of biotic and abiotic factors and how both sets of factors impact aquatic biomes. Students should work in pairs to do their research and design their solutions.

- Video: Hawaii and the Great Pacific Garbage Patch

TEI Attach Presentation

Evaluate `45–90 minutes`

Explain Question:
What can humans do to reduce the extinction rate?

Lesson Questions (LQ):
- Why does biodiversity matter?
- What is the relationship between human population size and rates of extinction?
- What are the main causes of biodiversity loss?
- What steps can humans take to prevent the loss of biodiversity?

Throughout instruction and the 5E learning cycle, you will have collected formative assessment data to drive the assignment of resources and experiences to students. Evaluate is intended to include summative assessment checks for proficiency. You can use the Explain and Lesson Questions for the concept as a summative assessment in a variety of ways such as these:

- Post each Lesson Question (LQ) in various locations in the classroom, and have small groups of students generate claim statements related to the Lesson Question (LQ). Other students can add to the claim, or refute the claim, during a gallery walk where they place additional pieces of evidence on each Lesson Question (LQ) poster.
- Assign small groups of students to each Lesson Question (LQ) and have the groups generate a poster, board, graphic, or piece of text that answers the question. Use a jigsaw approach and create a second set of groups that contain members from each Lesson Question (LQ) group to share their ideas.
- Ask students to return to their initial ideas for the Explain question and add additional details and evidence.

Encourage students to review the concept review and complete the Student Self-Check practice assessment prior to assigning the Summative Teacher Concept assessment.

- Student Review and Practice Assessment
- Teacher Concept Assessments

GLOSSARY

English ———— A ———— Español

abiotic non-living, physical components of the environment

abiótico componentes físicos no vivos del medio ambiente

abrasion a type of mechanical weathering caused by the scraping and scratching of rocks by loose particles that are transported over the rocks by wind, water, glaciers, etc.

abrasión tipo de desgaste mecánico originado por el rozamiento o fricción de partículas sueltas arrastradas por el viento, el agua, los glaciares, etc. sobre las rocas

activation energy the minimum amount of energy required to initiate a chemical reaction; written as Ea and measured in kilojoules

energía de activación cantidad mínima de energía necesaria para iniciar una reacción química; se representa como Ea y se mide en kilojulios

active transport movement of ions or molecules across a membrane, often against a concentration gradient; requires energy input; endocytosis is a type of active transport

transporte activo movimiento de iones o moléculas a través de una membrana, con frecuencia contra un gradiente de concentración; requiere un aporte de energía; la endocitosis es un tipo de transporte activo

adaptation a change in the function or structure of an organism that makes it better suited to its environment

adaptación cambio en la función o estructura de un organismo que lo hace más apto para su medio ambiente

adenine one of the four nitrogenous bases contained in a DNA molecule; adenine is complementary to guanine

adenina una de las cuatro bases nitrogenadas que forman parte de una molécula de ADN; la adenina es complementaria a la guanina

adenosine triphosphate (ATP) a chemical compound that is able to store and transport chemical energy within cells; also known as ATP

trifosfato de adenosina compuesto químico capaz de almacenar y transportar energía química en las células; también se conoce como ATP

adiabatic cooling occurs when the temperature of an air mass decreases as a result of the expansion of the air mass as it rises

enfriamiento adiabático se produce cuando la temperatura de una masa de aire disminuye como resultado de la expansión de dicha masa a medida que asciende

ADP ADP, adenosine diphosphate is a nucleotide that is involved in the transfers of energy in biological systems.

ADP adenosín difosfato es un nucleótido que interviene en la transferencia de energía en los sistemas biológicos.

aerobic respiration the form of in cell respiration that requires free oxygen

respiración aeróbica forma de respiración celular que requiere oxígeno

age pyramid a graphic depiction of the age and sex distribution of a population

pirámide de edad representación gráfica de la distribución por edad y sexo de una población

allele any of several possible forms of a gene	**alelo** cualesquiera de las varias formas posibles de un gen
allopatric speciation the evolution of a population into a new species when the population is separated from other members of the original species by a geographic barrier	**especiación alopátrica** evolución de una población dando lugar a especies nuevas cuando la población se separa de otros miembros de la especie original mediante una barrera geográfica
amino acid one of the 20 types of molecules that combine to form proteins	**aminoácido** uno de los 20 tipos de moléculas que se combinan para formar proteínas
anaerobic respiration the form of cell respiration that can take place in the absence of free oxygen	**respiración anaeróbica** tipo de respiración celular que puede tener lugar en ausencia de oxígeno
aneuploidy a condition in which one or more chromosomes are present in extra copies or are absent	**aneuploidía** condición en la cual existen copias adicionales de uno o más cromosomas o cromosomas ausentes
angular unconformity an unconformity in which younger sediments have been deposited on top of the eroded surface of tilted or folded older rock layers	**discordancia angular** discordancia en la cual los sedimentos más jóvenes se encuentran depositados sobre la superficie erosionada de capas de roca más antiguas dobladas o inclinadas
animal cell a form of a eukaryotic cell that is distinct from plant cells in that it does not have a rigid cell wall	**célula animal** tipo de célula eucariota que es diferente de la células vegetales; la célula animal no tiene una pared celular rígida
anthropogenic caused or originating from human beings	**antropogénico** causado por los seres humanos u originado en ellos
antibiotic a chemical that kills or inhibits growth of bacteria	**antibiótico** sustancia química que mata o inhibe el crecimiento de bacterias
anticodon a sequence of 3 nucleotides on a tRNA molecule that is complementary to a codon on mRNA	**anticodón** secuencia de 3 nucleótidos en una molécula de ARNt que es complementario de un codón en una molécula de ARNm
Archaea one of three domains of living organisms; single-celled prokaryotic micro-organisms that lack a nucleus; similar in size and shape to bacteria; often found in harsh environments	**Archaea** uno de los tres dominios de los seres vivos; microorganismos procariotas unicelulares que carecen de núcleo; semejantes en tamaño y forma a las bacterias; con frecuencia se encuentran en ambientes extremos

GLOSSARY

artery a blood vessel that carries blood away from the heart

arteria vaso sanguíneo que lleva sangre desde el corazón

artificial selection selective breeding of plants and animals by humans

selección artificial cría selectiva de plantas y animales llevada a cabo por los seres humanos

asexual reproduction the generation of offspring from a single parent that occurs without the fusion of gametes

reproducción asexual generación de descendencia a partir de un solo progenitor que se produce sin que haya fusión de gametos

asthenosphere the layer of soft but solid mobile rock found below the lithosphere. The asthenosphere begins about 100 km below Earth's surface and extends to a depth of about 350 km; the lower part of the upper mantle.

astenosfera capa de roca capaz de moverse, blanda pero sólida, que se encuentra bajo la litosfera. La astenosfera comienza aproximadamente a 100 km bajo la superficie de la Tierra y se extiende hasta una profundidad de unos 350 km; parte inferior del manto superior.

atmosphere the layers of gases that surround a planet

atmósfera capas de gases que rodean un planeta

autosomal dominant disorder a disorder in which a person needs only to inherit an abnormal gene from one parent in order to develop the disease

trastorno autosómico dominante trastorno en el cual una persona necesita solo heredar un gen anormal de uno de los progenitores para desarrollar la enfermedad

autosomal recessive disorder a disorder in which a person must inherit two copies of an abnormal gene—one from each parent—in order for the disease or trait to develop

trastorno autosómico recesivo trastorno en el cual una persona debe heredar dos copias de un gen anormal (uno de cada progenitor) para desarrollar la enfermedad o el rasgo

autotroph organism that can create its own food from simple molecules

autótrofo organismo que puede producir su propio alimento a partir de moléculas simples

B

bacteria one of three domains of living organisms; single-celled prokaryotic organisms that lack an organized nucleus; similar in size and shape to Archaea; can be found in nearly every habitat on Earth

bacteria uno de los tres dominios de los seres vivos; procariotas unicelulares que carecen de núcleo organizado; semejantes en tamaño y forma a las Archaea; se pueden encontrar en casi cualquier hábitat en la Tierra

bilateral symmetry a characteristic of organisms that can be divided in half along a midline yielding similar left and right halves

simetría bilateral característica de los organismos según la cual estos pueden dividirse a la mitad a lo largo de una línea central, dando lugar a dos mitades similares, una derecha y otra izquierda

binary fission a method of asexual reproduction in which single-celled organisms like bacteria reproduce by replicating their DNA and splitting in half

fisión binaria método de reproducción asexual en el cual organismos unicelulares como las bacterias se reproducen replicando su ADN y dividiéndose en dos mitades

biodiversity the variety of species that exist in an environment

biodiversidad variedad de especies que existen en un medio ambiente

bio-fuel a type of renewable energy source created from organisms

biocombustible tipo de energía renovable producida a partir de organismos

biogeochemistry a combination of disciplines that studies the biochemistry of flora and fauna in a geographical context

biogeoquímica combinación de disciplinas que estudia la bioquímica de la flora y la fauna en un contexto geográfico

biogeography the study of the environmental distribution of organisms

biogeografía estudio de la distribución de los organismos en los medioambientes

biomagnification the increase in or process of increasing the concentration of a substance (usually a toxin) in living tissue at each level of the food chain

biomagnificación aumento o proceso de aumento de la concentración de una sustancia (por lo general, una toxina) en tejidos vivos en cada nivel de la cadena alimentaria

biomass the mass of dried living matter in a given area or volume of habitat

biomasa masa de materia orgánica seca que se encuentra en un área dada o en un volumen de hábitat

biome a major ecological community such as grassland, tropical rain forest, or desert

bioma gran comunidad ecológica, como una pradera, una selva tropical o un desierto

biosphere the part of Earth in which living organisms are known to exist

biósfera parte de la Tierra en la que existen los organismos vivos

biotic living, physical components of the environment

biótico vivo, componentes físicos del medio ambiente

blood tissue that brings oxygen and nutrients to the cells of the body and removes wastes from the cells of the body

sangre tejido que lleva oxígeno y nutrientes a las células del cuerpo y remueve los desechos de las células del cuerpo

bottleneck effect a limited variety of alleles in a population due to a dramatic decrease in population size

efecto cuello de botella limitada variedad de alelos en una población debida a una disminución drástica del tamaño de la población

brain stem the part of the brain that connects to the spinal cord

tronco encefálico parte del cerebro que está conectada con la médula espinal

GLOSSARY

C

Calvin cycle the carbon-fixing reaction in photosynthesis, also known as the dark or light-independent reactions; the Calvin cycle uses ATP and NADPH produced in the light reactions to incorporate the carbon from carbon dioxide into carbohydrates

ciclo de Calvin reacción de fijación del carbono en la fotosíntesis, también conocida como reacciones independientes clara y oscura; en el ciclo de Calvin se usa ATP y NADPH producidos en las reacciones claras para incorporar el carbono procedente del dióxido de carbono y producir hidratos de carbono

Cambrian explosion a period of rapid growth in life's diversity from 540 to 485 million years ago

explosión cámbrica período de crecimiento rápido en la biodiversidad hace 540 a 485 millones de años

cancer cells cells that grow and divide continuously at an unregulated pace

células cancerosas células que crecen y se dividen continuamente a un ritmo descontrolado

carbohydrate one of four major classes of organic compounds in living cells and an important source of nutritional energy; includes simple sugars, and more complex sugars such as starch sugars or many sugars; can also serve as a structural molecule

hidrato de carbono una de las cuatro clases principales de compuestos orgánicos presentes en las células vivas e importante fuente de energía nutricional; este tipo de compuesto incluye azúcares simples, azúcares más complejos como el almidón y muchos azúcares; también puede funcionar como molécula estructural

carbon cycle a natural cycle in which carbon compounds, mainly carbon dioxide, are incorporated into living tissue through photosynthesis and returned to the atmosphere by respiration, decay of dead organisms, and the burning of fossil fuels

ciclo del carbono ciclo natural en el cual los compuestos de carbono, principalmente el dióxido de carbono, se incorporan en los tejidos vivos a través de la fotosíntesis y regresan a la atmósfera por medio de la respiración, la desintegración de organismos muertos y la quema de combustibles fósiles

carbon reservoir a component in the carbon cycle in which carbon is stored, such as the atmosphere, oceans, biosphere, or lithosphere. Reservoirs can serve as carbon sinks or carbon sources.

depósito de carbono componente en el ciclo del carbono en el cual se almacena el carbono, como la atmósfera, los océanos, la biosfera o la litosfera. Los depósitos pueden servir como sumideros de carbono o fuentes de carbono.

carbon the chemical element with the atomic number 6

carbono elemento químico con número atómico 6

carbonate an ion composed of carbon and oxygen in a 1:3 ratio with a positive 2 charge (CO_3^{+2}); a general term for minerals and other substances that contain carbonate ions as part of their chemical structure, such as calcite, aragonite, and dolomite

carbonato ión compuesto por carbono y oxígeno en una proporción 1:3 con 2 cargas positivas (CO_3^{+2}); término general para minerales y otras sustancias que contienen iones de carbonato como parte de su estructura química, como la calcita, el aragonito y la dolomita

carnivore an animal that eats only other animals

carnívoro animal que se alimenta solo de otros animales

carrying capacity indicates the greatest number of any species that can indefinitely exist within a specific habitat without threatening the existence of other species also living in the habitat

capacidad de carga indica el mayor número de individuos de cualquier especie que puede existir indefinidamente en un hábitat específico sin poner en riesgo la existencia de otras especies que también viven en ese hábitat

catalyst substance that increases the rate of a chemical reaction by lowering the amount of energy needed for the reaction to occur, but is not changed by the reaction

catalizador sustancia que aumenta la velocidad de una reacción química al disminuir la cantidad de energía necesaria para que la reacción se produzca, pero no es modificada por la reacción

cell cycle the sequence of phases that a eukaryotic cell progresses through beginning with its origin in the division of a parent cell and ending with its own division

ciclo celular secuencia de fases por las que atraviesa una célula eucariota, desde el principio con su origen en la división de una célula progenitora hasta terminar con su propia división

cell membrane a biological membrane, also called the plasma membrane, that surrounds a cell and selectively controls which substances can enter or leave the cell

membrana celular membrana biológica, también llamada membrana plasmática, que rodea a la célula y controla, de manera selectiva, qué sustancias pueden entrar en la célula o salir de ella

cell theory the theory that (1) all organisms are made out of one or more cells; (2) cells are the smallest units of life; (3) cells come from pre-existing cells via cell division

teoría celular teoría de que (1) todos los organismos están compuestos por una o más células; (2) las células son las unidades de vida más pequeñas; (3) las células provienen de células preexistentes mediante división celular

cellular division the process in which a cell splits in two; part of the larger cell cycle. In eukaryotes, there are two kinds of cell division: mitosis and meiosis. Prokaryotes divide by the process of binary fission.

división celular proceso en el cual una célula se divide en dos; parte del ciclo celular más largo. En las eucariotas hay dos tipos de división celular: mitosis y meiosis. Las procariotas se dividen mediante el proceso de fisión binaria.

GLOSSARY

cellular respiration the process that occurs when the chemical energy of "food" molecules, carbohydrates, fats, and proteins, is released and partially captured in the form of adenosine triphosphate (ATP)

respiración celular proceso que ocurre cuando la energía química de las moléculas de "comida," hidratos de carbono, grasas y proteínas, se libera y es capturado parcialmente en forma de trifosfato de adenosina (ATP)

cellulose an insoluble polymer that is the main substance in plant cell walls.

celulosa polímero insoluble que es la sustancia principal de las paredes de las células vegetales

chaparral biome an ecological community characterized by a hot, dry summer and mild, wet winters, scrub vegetation, flat plains, rocky hills, and mountain slopes

bioma de chaparral comunidad ecológica caracterizada por veranos secos y cálidos e inviernos húmedos, vegetación de arbustos, llanuras, colinas rocosas y montañas de pendientes suaves

Charles Darwin (1809–1882) British scientist, laid the foundation of modern with his concept that all species have developed from a common ancestor through the process of natural selection.

Charles Darwin (1809–1882) Científico británico, sentó la base de la biología moderna con su concepto de que todas las especies se han desarrollado a partir de un ancestro común a través del proceso de selección natural.

chemical energy the energy that is stored in the bonds between atoms

energía química energía que se encuentra almacenada en los enlaces entre átomos

chemical weathering changes to rocks and minerals on Earth's surface that are caused by chemical reactions

meteorización química cambios en las rocas y minerales de la superficie de la Tierra causados por reacciones químicas

chlorophyll a green pigment found in photosynthetic organisms

clorofila pigmento verde que se encuentra en los organismos fotosintéticos

chloroplast an organelle in plant and algae cells that converts energy from sunlight into chemical energy that the plant can use

cloroplasto orgánulo en las células de las plantas y las algas que convierte la energía de la luz solar en energía química que la planta puede usar

chromatid a replicated chromosome, joined by the centromere to a copy of the chromosome

cromátido cromosoma replicado, unido por el centrómero a una copia del cromosoma

chromatin a macromolecular complex enabling efficient packaging of DNA

cromatina complejo macromolecular que permite la envoltura eficiente del ADN

chromosomal abnormalities errors in the structure or number of chromosomes

anormalidades cromosomáticas errores en la estructura o en el número de cromosomas

chromosome Chromosomes are the intracellular structures made of the cell's double-stranded DNA genome packaged bound to DNA-binding proteins.

climate the current or past long-term weather conditions characteristic of a region or the entire Earth

coal a solid fossil fuel composed primarily of carbon that forms from decomposed plant materials

codominant allele a version of a gene that is co-expressed with another allele in the heterozygote

codon the portion of a DNA or mRNA molecule that encodes for a specific amino acid or marks the starting or stopping of protein production

coenzyme an organic molecule that binds to an enzyme's active site to enable catalysis

commensalism a relationship between two species of a plant, animal, or fungus in which one lives with, on, or in the other without damage to either

community a group of different populations that live together and interact in an environment

comparative morphology comparing the structures and features of organisms to determine evolutionary relationships

competition the interaction between organisms or species that use the same resources in which the health of one is negatively affected by the presence of the other

condensation the process by which a gas changes into a liquid

cromosoma Los cromosomas son estructuras intracelulares hechas de paquetes de genoma de dobles hebras de ADN de la célula unido a proteínas de enlace de ADN.

clima condiciones atmosféricas, actuales o pasadas, durante un largo periodo de tiempo, características de una región o de toda la Tierra

carbón combustible fósil compuesto principalmente por carbón que se forma a partir de materia vegetal descompuesta

alelo codominante versión de un gen que está co-expresada con otro alelo en el heterocigoto

codon parte de la molécula del ADN o del ARNm que codifica para un aminoácido específico o que marca el comienzo o el cese de la producción de proteína

coenzima molécula orgánica que se une al sitio activo de una enzima para permitir la catálisis

comensalismo relación entre dos especies de una planta, animal u hongos en la cual uno vive en el otro sin dañarlo

comunidad grupo de poblaciones diferentes que viven juntas e interactúan en un medio ambiente

morfología comparativa comparación de estructuras y características de organismos para determinar relaciones evolutivas

competencia interacción entre organismos o especies que usan los mismos recursos en la cual la salud de uno se ve afectada de manera negativa por la presencia del otro

condensación proceso mediante el cual un gas cambia a estado líquido

GLOSSARY

conservation the act of preserving natural resources, the environment, or other valuable commodities

conservación acto de preservar los recursos naturales, el medio ambiente u otras cosas valiosas

continental crust the rocks of Earth's crust that make up the base of the continents, ranging in thickness from about 35 km to 60 km under mountain ranges. Continental crust is generally less dense than oceanic crust.

corteza continental rocas de la corteza terrestre que constituyen la base de los continentes, su espesor va desde alrededor de 35 km hasta 60 km bajo cadenas montañosas. Por lo general la corteza continental es menos densa que la corteza oceánica.

continental drift the movement of Earth's continents relative to each other

deriva continental movimiento de los continentes de la Tierra respecto los unos de los otros

convection the transfer of heat from one place to another caused by movement of molecules

convección transferencia de calor de un lugar a otro producido por el movimiento de moléculas

convergent boundary a tectonic plate boundary at which two tectonic plates move toward each other, causing collisions and subduction zones

límite convergente límite de una placa tectónica en el cual dos placas se mueven una hacia otra y producen colisiones y zonas de subducción

convergent evolution a process by which two unrelated organisms will share similar features due to the similar pressures of natural selection

evolución congervente proceso por el que dos organismos no relacionados comparten características similares debido a las presiones similares de la selección natural

core the innermost layer of Earth, comprised of the liquid outer core and solid inner core; consists mainly of iron and nickel

núcleo capa más interna de la Tierra, consta del núcleo externo líquido y el núcleo interno sólido; constituido principalmente por hierro y níquel

crust the outermost rocky layer of a rocky planet or moon, which is chemically distinct from an underlying mantle

corteza capa rocosa más externa de un planeta o luna rocoso que se distingue químicamente del manto subyacente

cryosphere all of the solid water on Earth, including snow and ice

criósfera toda el agua sólida de la Tierra, incluyendo nieve y hielo

cyanobacteria phylum of photosynthetic bacteria, also called "blue-green algae"

cianobacterias filo de bacterias fotosintéticas, también llamadas "algas azules verdosas"

cytosine one of the four nitrogenous bases contained in a DNA molecule. Cytosine is complementary to guanine.

citosina una de las cuatro bases nitrogenadas que forman parte de una molécula de ADN. La citosina es complementaria a la guanina.

cytoskeleton a network of protein fibers found in eukaryotic cells

cytosol the fluid part of a cell

D

decomposer organisms which carry out the process of decomposition by breaking down dead or decaying organisms

density-dependent factors environmental factors that are affected by the population density, or the number of organisms in a specific area

density-independent factors environmental factors that are not affected by the population size or density

deposition (sedimentary) occurs when eroded sediments are dropped in another location, ending the process of erosion

desert biome a major ecological community defined by hot, arid conditions and extremely low rainfall

detritivore an organism that gains nutrition by consuming decomposing plant and animal material as well as organic fecal matter

detritus material resulting from erosion, waste or debris resulting from biological decomposition

differential weathering the chemical or physical weathering of rocks that occurs at different rates, producing an uneven surface in the rock layers

dihybrid cross a cross between two organisms that are each heterozygous for two traits of interest

citoesqueleto red de fibras de proteínas que se halla en las células eucariotas

citosol parte líquida de una célula

descomponedor organismo que lleva a cabo el proceso de descomposición mediante la desintegración de los organismos muertos

factores dependientes de la densidad factores medioambientales que se ven afectados por la densidad de población o número de organismos en un área específica

factores independientes de la densidad factores medioambientales que no se ven afectados por el tamaño de la población ni por su densidad

deposición (sedimentaria) ocurre cuando los sedimentos erosionados se trasladan a otra ubicación y se quedan ahí, poniendo así fin al proceso de erosión

bioma de desierto gran comunidad ecológica con condiciones cálidas, áridas y cantidades de lluvia extremadamente bajas

detritívoro organismo que se nutre de materia animal y vegetal en descomposición y materias fecales orgánicas

detritus material producto de la erosión, desecho o residuo producto de la descomposición biológica

meteorización diferencial meteorización química o física de las rocas que ocurre a diferentes velocidades, produciendo una superficie irregular en las capas rocosas

cruce dihíbrido cruce entre dos organismos heterocigóticos para dos rasgos en los que se tiene interés

GLOSSARY

diploid cell a cell that has two complete sets of chromosomes and is designated as 2n

célula diploide célula que tiene dos grupos completos de cromosomas y se designa como 2n

directional selection a type of natural selection where selective pressures on a species favors one phenotype to be selected

selección direccional tipo de selección natural en la que las presiones selectivas sobre una especie favorece la selección de un fenotipo

discharge (stream) flow of water (as from a stream, river or pipe)

descarga (corriente) flujo de agua (de un arroyo, río o tubería)

disconformity an unconformity in which the rock layers are parallel

disconformidad discordancia en la cual las capas rocosas son paralelas

disruptive selection a type of natural selection where selective pressures on a species favor extremes at both ends of the phenotypic range of traits

selección disruptiva tipo de selección natural en la que las presiones selectivas sobre una especie favorece ambos extremos del rango fenotípico de los rasgos

divergent boundary a tectonic plate boundary at which two tectonic plates move away from each other

límite divergente límite de una placa tectónica en el cual dos placas tectónicas se mueven separándose una de otra

DNA deoxyribonucleic acid; a molecule found in cells that carries genetic information to be passed from parents to offspring during reproduction

ADN ácido desoxirribonucleico; molécula que se encuentra en las células que contienen información genética que puede pasar de progenitores a descendientes durante la reproducción

dominant allele an allele that is fully expressed in the phenotype of a heterozygote

alelo dominante alelo que está plenamente expresado en el fenotipo de un heterocigoto

double helix the shape of a DNA molecule; two spiral strands wrapped around each other

hélice doble forma de una molécula de ADN; dos hebras espirales enrolladas una alrededor de la otra

E

Earth the third planet from the sun; the planet on which humans and other organisms live

Tierra tercer planeta a partir del Sol; planeta en el cual vivimos los seres humanos y otros organismos

ecosystem all the living and nonliving things in an area that interact with each other

ecosistema todos los seres vivos y los elementos no vivos en un área que interaccionan entre sí

El Niño 1. a short period of abnormally warm temperatures across the globe caused by a band of warm ocean water that occasionally develops off the western coast of South America 2. the ocean current responsible for El Niño events

electron a negatively charged subatomic particle that exists in various energy levels outside the nucleus of an atom

electron transport chain a series of proteins in the mitochondrial membrane which transfer electrons to oxygen in a step-wise process which produces ATP

elevation the vertical distance above sea level of a point on Earth's surface

endergonic reaction a chemical reaction in which energy is absorbed because the standard change in free energy is positive

endocrine system an internal system composed of several glands that secrete hormones directly into the blood that are related to growth, emotional responses, and sleep cycles among other things

endocytosis a process in which cells take in materials by wrapping a section of plasma membrane around those materials and bringing them into the cell

endothermic a reaction that absorbs heat from the surrounding area

energy the ability to do work or cause change; can be stored in chemicals found in food and released to the organism to do work

energy pyramid a model that shows the available amount of energy in each trophic layer in an ecosystem

El Niño 1. periodo corto de temperaturas anormalmente cálidas en todo el planeta causadas por una banda de aguas oceánicas cálidas que se desarrolla de vez en cuando en la costa oeste de América del Sur 2. corriente oceánica responsable de El Niño

electrón partícula subatómica con carga negativa que existe en varios niveles de energía alrededor del núcleo de un átomo

cadena de transporte de electrones serie de proteínas en la membrana mitocondrial la cual transfiere electrones al oxígeno en un paso del proceso de producción de ATP

elevación distancia vertical sobre el nivel del mar desde un punto de la superficie terrestre

reacción endergónica reacción química en la cual la energía es absorbida debido a que la carga estándar en la energía libre es positiva

sistema endocrino sistema interno compuesto por varias glándulas que segregan hormonas directamente en la sangre; las hormonas están relacionadas con el crecimiento, las repuestas emocionales y los ciclos del sueño, entre otras cosas

endocitosis proceso en el cual las células toman materiales envolviendo una parte de la membrana plasmática alrededor de estos materiales e introduciéndolos en la célula

endotérmico reacción que absorbe calor de su entorno

energía capacidad de realizar trabajo o producir un cambio; puede almacenarse en sustancias químicas que se encuentran en los alimentos y liberarse al organismo para realizar trabajo

pirámide de energía modelo que muestra la cantidad disponible de energía en cada nivel trófico de un ecosistema

GLOSSARY

enzyme proteins that catalyze specific chemical reactions in organisms by lowering the activation energy of the reaction

enzima proteína que cataliza reacciones químicas específicas en los organismos al disminuir la energía de activación de la reacción

episodic speciation the evolution of many new species in a relatively short period of time

especiación episódica evolución de muchas especies nuevas en un periodo de tiempo relativamente corto

erosion the process by which wind, water, ice, gravity, or other natural forces move sediment over Earth's surface

erosión proceso en el cual el viento, el agua, el hielo, la gravedad u otras fuerzas naturales desplazan sedimentos sobre la superficie de la Tierra

eukaryotic cell a type of cell with a nucleus enclosed by a membrane as well as membrane-enclosed organelles

célula eucariota tipo de célula que tiene un núcleo dentro de una membrana y orgánulos, también dentro de membranas

eustatic (change) alteration of global sea levels

austático (cambio) alteración del nivel del mar global

evaporation the process in which matter changes from a liquid to a gas

evaporación proceso en el cual la materia cambia de estado líquido a gaseoso

evolutionary tree diagram a diagram that depicts the evolutionary relationships between organisms

árbol filogenético diagrama que muestra las relaciones evolutivas entre los organismos

exergonic reaction a chemical reaction in which energy is released in the form of heat because the change in free energy is negative

reacción exergónica reacción química en la cual la energía es liberada en forma de calor debido a que la carga estándar en la energía libre es negativa

exfoliation a mechanical weathering process in which thin layers of rock on the outer surfaces of outcrops or other rock features break off, often creating dome-shaped patterns

exfoliación proceso de desintegración mecánica en el cual se desprenden finas capas de roca de las superficies más exteriores de afloramientos o desintegración de otras estructuras rocosas, para crear patrones con forma de domo

exocytosis a process in which cells expel materials within a vesicle by fusing the vesicle to the plasma membrane

exocitosis proceso en el cual las células expulsan materiales del interior de una vesícula fusionando esta con la membrana plasmática

exotic species a species either deliberately or accidentally introduced to a range that it is not native to

especie exótica especie que es introducida de manera deliberada o por accidente en un entorno en el que no es nativa

exponential growth growth of biological organisms which goes on unhindered when resources are unlimited

extinction the permanent loss of a population or species

extinction rate a measure of the frequency at which organisms become extinct

F

facilitated diffusion a process in which chemicals that are unable to cross the membrane directly diffuse across the membrane through transport proteins. These proteins facilitate (or "help") the process of diffusion.

fat a triester of triglycerol and fatty acids

feedback mechanism the process in which part of the output of a system is returned to the input to regulate further output

fermentation conversion of carbohydrates into alcohols

fertilization the process in which two gametes, such as an egg and sperm, unite to form a new organism, or zygote

food chain a model that shows one set of feeding relationships among living things

food web a model that shows many different feeding relationships among living things in a given area

fossil the preserved remains of an organism, or traces of an organism such as a mark or print left by an animal

crecimiento exponencial crecimiento de organismos biológicos sin obstáculos cuando los recursos son ilimitados

extinción pérdida permanente de una población o especie

tasa de extinción medida de la frecuencia a la que se extinguen los organismos

difusión facilitada proceso en el cual las sustancias químicas que no son capaces de cruzar directamente la membrana se difunden a través de la membrana mediante proteínas transportadoras. Estas proteínas facilitan (o "ayudan") el proceso de difusión.

grasa triéster de triglicerol y ácidos grasos

mecanismo de retroalimentación proceso en el que parte del producto de un sistema regresa al sistema para regular más productos

fermentación conversión de hidratos de carbono en alcoholes

fertilización proceso en el cual dos gametos, como un óvulo y un espermatozoide, se unen para formar un nuevo organismo, o cigoto

cadena alimentaria modelo que muestra un conjunto de relaciones alimentarias entre seres vivos

red alimentaria modelo que muestra relaciones alimentarias muy variadas entre los seres vivos de un área determinada

fósil restos conservados de un organismo o de huellas de un organismo como marcas o huellas dejadas por un animal

GLOSSARY

fossil fuel a nonrenewable resource formed from organic carbon due to the compression and partial decomposition of organisms

combustible fósil recurso no renovable formado a partir del carbono orgánico originado por la compresión y la descomposición parcial de organismos

founder effect the loss of genetic diversity that occurs when a small number of individuals from a large population of a species establish a new population

efecto fundador pérdida de diversidad genética que ocurre cuando un pequeño número de individuos de una gran población de una especie establece una nueva población

fresh water water with a low salt concentration—usually 1% or less

agua dulce agua con una concentración baja de sal; por lo general, de menos del 1%

freshwater biome a major ecological community defined by freshwater regions that contain low salt content and experience average precipitation

bioma de agua dulce gran comunidad ecológica definida por regiones de agua dulce que contiene poca cantidad de sal y recibe precipitaciones promedio

fructose a naturally occurring monosaccharide

fructosa monosacárido natural

G

gabbro a black, mafic, coarse-grained intrusive igneous rock; the intrusive equivalent of basalt

gabro roca ígnea, intrusiva, máfica, de grano grueso y color negro; roca intrusiva equivalente del basalto

gene expression the result of coding information determined by DNA

expresión génica resultado de la codificación de información determinada por el ADN

gene therapy the introduction of cloned genes into living cells to replace an abnormal, disease-causing gene

terapia génica introducción de genes clonados en células vivas para sustituir un gen anormal causante de una enfermedad

genetic disorder an illness caused by abnormalities in genes or chromosomes

trastorno genético enfermedad producida por anormalidades en genes o cromosomas

genetic drift a random change in the allele frequency of a population over successive generations

deriva genética cambio aleatorio en la frecuencia de los alelos de una población en generaciones sucesivas

genetic testing tests that are used to predict whether a person is at risk of developing or is susceptible to a particular disease

examen genético examen que se realiza para predecir si una persona es susceptible a cierta enfermedad o si tiene riesgo de desarrollarla

genetic variation range of differences in DNA within a population or species organisms

variación genética rango de diferencias en el ADN dentro de una población o especie de organismos

genome total of all DNA sequences in a cell or organism

genotype the genetic code passed down from parents to offspring that determines all physical and physiological traits, also called phenotypes

geologic time time measured on the scale of Earth's 4.56-billion-year history, as determined by the rock and fossil records

geology the study of Earth through the study of rocks, minerals, water, and other Earth materials, as well as through seismic waves and other natural phenomena

geosphere the combination of Earth's inner and outer core, mantle, and crust

geothermal energy a natural, renewable energy resource produced by Earth's naturally occurring heat, steam, and hot water

glacial varve alternating light- and dark-colored layers of sediment, deposited periodically in a glacial lake, that can be used to date annual, cyclical, or seasonal changes

glacier a large mass of ice resting on, or overlapping, a land surface

global warming the slow increase of Earth's average global atmospheric temperature due to climatic change

glucose a carbohydrate; produced by photosynthesis; primary source of energy for some plant and animal cells

glycolysis the breakdown of glucose through a series of biochemical reactions produces two molecules each of ATP, pyruvate, and NADH

genoma total de todas las secuencias de ADN en una célula u organismo

genotipo código genético que se transmite de los progenitores a su descendencia y que determina todos los rasgos físicos y psicológicos, también denominados fenotipos

tiempo geológico tiempo medido en la escala de la historia de la Tierra de 4.56 mil millones de años, determinada por el estudio de rocas y registros fósiles

geología estudio de la Tierra a través del estudio de las rocas, minerales, aguas y otros materiales terrestres y mediante el estudio de ondas sísmicas y otros fenómenos naturales

geosfera combinación de los núcleos interno y externo de la Tierra, el manto y la corteza

energía geotérmica recurso energético renovable producido de manera natural por la Tierra, presente en forma de calor, vapor y agua caliente

varva glacial capas alternas de sedimentos, en colores claros y oscuros, que se depositan periódicamente en un lago glaciar, pueden usarse para datar cambios anuales, cíclicos o estacionales

glaciar masa grande de hielo que se encuentra sobre una superficie de tierra o se superpone a ella

calentamiento global lento aumento de la temperatura atmosférica promedio de la Tierra debido al cambio climático

glucosa hidrato de carbono; producido por fotosíntesis; fuente primaria de energía para algunas células de animales y plantas

glicólisis desintegración de la glucosa mediante una serie de reacciones bioquímicas produce dos moléculas de ATP, piruvato y NADH

GLOSSARY

grassland biome a major ecological community characterized by extensive areas grasses, flowers, and herbs and an erratic precipitation that is enough to support such vegetation, but very few trees

bioma de pradera gran comunidad ecológica caracterizada por extensas áreas de pasto, flores y hierbas; recibe precipitaciones variables suficientes para sostener la vida de este tipo de vegetación, pero muy pocos árboles

gravity a force that exists between any two objects that have mass and that pulls the objects together. The greater the mass of an object, the greater its gravitational pull.

gravedad fuerza entre dos objetos cualesquiera que tienen masa y que atrae uno hacia el otro. Cuanto mayor es la masa de un objeto, mayor es la atracción gravitacional.

greenhouse gas a gas, usually carbon-based, that contributes to global warming through the greenhouse effect, which prevents the escape of radiant heat from Earth's atmosphere

gas invernadero gas, por lo general a base de carbono, que contribuye al calentamiento global mediante el efecto invernadero, el cual impide que el calor radiante salga de la atmósfera terrestre

Gregor Mendel (1822–1884) Austrian scientist and monk who is considered the founder of the science of genetics. His quantitative analysis of the inheritance of certain traits in pea plants allowed him to deduce two fundamental principles known as the laws of Mendelian

Gregor Mendel (1822–1884) Científico y monje austriaco considerado el fundador de la ciencia de la genética. Sus análisis cuantitativos sobre la herencia de ciertos rasgos en plantas de arvejas le permitieron deducir dos principios fundamentales conocidos como leyes de Mendel

groundwater water stored below Earth's surface in soil and rock layers

agua subterránea agua almacenada bajo la superficie de la Tierra en capas de suelo y roca

guanine one of the four nitrogenous bases contained in a DNA molecule; guanine is complementary to cytosine

guanina una de las cuatro bases nitrogenadas que forman parte de una molécula de ADN; la guanina es complementaria a la citosina

H

haploid cell a cell that contains one complete set of chromosomes and is designated as 1n

célula haploide célula que tiene un grupo completo de cromosomas y se designa como 1n

Hardy-Weinberg equation $p^2 + 2pq + q^2 = 1$; The frequencies of two alleles, p and q, in a population will remain stable if no selective pressures are acting on a sufficiently large population.

ecuación de Hardy-Weinberg $p^2 + 2pq + q^2 = 1$; Las frecuencias de dos alelos, p y q, en una población permanecerán estables si no actúan presiones selectivas en una población suficientemente grande.

heart the muscular organ of an animal that pumps blood throughout the body

heat energy a form of energy that transfers between particles in a substance or system through kinetic energy transfer

herbivore an animal that consumes only plants

heterotroph an organism that cannot synthesize its own food from simple substances

heterozygous an organism that has two different alleles for a trait

homeostasis the ability of the internal systems of an organism to maintain normal chemical balance, despite changing external conditions

homology shared ancestry of biological structures or molecules

homozygous an organism that has a pair of identical alleles for a trait

hormone signaling molecule produced by glands and targeting organs to maintain homeostasis

hydroelectric energy electricity generated by moving water flowing over a turbine

hydrogen bond a weak type of chemical link between a negatively charged atom and a hydrogen atom that is bonded to another negatively charged atom

hydrolysis a chemical weathering process resulting from the reaction between the ions of water (H^+ and OH^-) and the ions of a mineral

hydrosphere all of the water on, under, and above Earth

corazón órgano muscular de un animal que bombea sangre a través del cuerpo

energía calorífica forma de energía que se transfiere entre partículas en una sustancia o en un sistema por medio de transferencia de energía cinética

herbívoro animal que se alimenta solo de plantas

heterótrofo organismo que no puede sintetizar su propio alimento a partir de sustancias simples

heterocigótico organismo que tiene dos alelos diferentes para un rasgo

homeostasis capacidad de los sistemas internos de un organismo para mantener un equilibrio químico normal a pesar de los cambios en las condiciones externas

homología ascendencia compartida de estructuras o moléculas biológicas

homocigótico organismo que tiene un par de alelos idénticos para un rasgo

hormona molécula de señalización producida por una glándula que se dirige hacia los órganos para mantener la homeostasis

energía hidroeléctrica electricidad generada por agua que discurre rápidamente sobre una turbina

enlace de hidrógeno tipo de enlace químico débil entre un átomo con carga negativa y un átomo de hidrógeno que está enlazado con otro átomo con carga negativa

hidrólisis proceso de meteorización química que es el resultado de la reacción entre los iones de agua (H^+ y OH^-) y los iones de un mineral

hidrosfera todo el agua existente en la Tierra, bajo ella o sobre ella

GLOSSARY

I

ice cap   glaciated areas centered around geographic poles

casquete de hielo   áreas glaciales que rodean a los polos geográficos

ice core   a sample of ice taken by a hollow tube from a glacier or other large ice body

núcleo de hielo   muestra de hielo que se toma de un glaciar o de otro gran cuerpo de hielo mediante un tubo hueco

ice wedging   a type of mechanical weathering in which water gets into cracks or joints in a rock, then freezes and expands, pushing the rock apart

cuña de hielo   tipo de meteorización mecánica en la cual el agua penetra en grietas o juntas en una roca, luego se congela y expande separando trozos de la roca

immune system   the body's natural defense against infection and illness

sistema inmunitario   defensas naturales del cuerpo contra infecciones y enfermedades

incomplete dominance   a form of inheritance in which the phenotype of a heterozygote is intermediate between the phenotypes of individuals who are homozygous for each allele

dominancia incompleta   forma de herencia en la cual el fenotipo de un heterocigoto es intermedio entre los fenotipos de individuos que son homocigotos para cada alelo

induced mutation   changes in DNA caused by mutagens such as chemicals or radiation

mutación inducida   cambios en el ADN originados por mutágenos como sustancias químicas y radiaciones

Industrial Revolution   period in Western history characterized by rapid advances of technology and production

Revolución Industrial   período de la historia occidental que se caracteriza por adelantos rápidos de la tecnología y la producción

inner core   the solid, inner portion of Earth's core, composed of an alloy of iron, nickel, and other heavy elements; rotates within the liquid outer core

núcleo interno   parte interna y sólida de la Tierra, compuesta por una aleación de hierro, níquel y otros elementos pesados; rota dentro del núcleo externo líquido

inorganic molecule   any molecule that is not considered to be of a biological origin

molécula inorgánica   cualquier molécula que no se considera de origen biológico

integumentary system   a body system consisting of the skin, feathers, and other outer coverings of an animal that help protect the animal from external damage

sistema tegumentario   sistema corporal que consiste en la piel, plumas y otras cubiertas exteriores de los animales que ayudan a proteger al animal de los daños del exterior

invertebrate   an animal that does not have a backbone

invertebrado   animal que no tiene columna vertebral

ion channel regulates the flow of ions across the membrane in all cells; the flow of ions through an ion channel does not require energy

ion pump also known as an ion transporter; consumes energy to move ions across cellular membranes

K

karst limestone formation with caverns and ravines

keystone species within the ecological community, this species has a critical role in maintaining the structure of the community

kinetic energy the energy an object has due to its motion

L

lactose a disaccharide formed from glucose and galactose

latitude angular distance north and south of the equator

Law of Independent Assortment one of Mendel's laws that states each pair of alleles segregates independently of one another during the formation of gametes

Law of Segregation one of Mendel's laws that states that two copies of a gene will segregate so that each gamete receives only one copy

lipid a group of organic compounds that are not soluble in water, but can be dissolved by other nonpolar solvents

liquid a state of matter with a defined volume but no defined shape and whose molecules roll past each other

canal iónico regula el flujo de iones a través de la membrana en todas las células; el flujo de iones a través de un canal iónico no requiere energía

bomba de iones también conocida como transportador de iones; consume energía para mover los iones a través de membranas celulares

karst formación de piedra caliza con cavernas y barrancos

especies clave dentro de la comunidad ecológica, estas especies tienen un papel crucial en el mantenimiento de la estructura de la comunidad

energía cinética energía que tiene un objeto debido a su movimiento

lactosa disacárido formado por glucosa y galactosa

latitud distancia angular al norte y sur del ecuador

ley de la recombinación independiente una de las leyes de Mendel que establece que cada par de alelos se segregan independientemente uno del otro durante la formación de gametos

ley de la segregación una de las leyes de Mendel que establece que dos copias de un gen se segregan para que cada gameto reciba una sola copia

lípidos grupo de compuestos orgánicos que no son solubles en agua, pero que pueden disolverse en otros solventes no polares

líquido estado de la materia con un volumen definido pero no forma definida y cuyas moléculas se deslizan unas sobre otras

GLOSSARY

lithosphere the part of Earth which is composed mostly of rocks; the crust and outer mantle

litosfera parte de la Tierra compuesta principalmente por rocas; corteza y manto exterior

logarithmic a scale in which the logarithm of a number with a given base value is the exponent that the base must be raised to in order to obtain the number

logarítmica escala en la cual el logaritmo de un número con una base dada es el exponente al cual hay que elevar la base para obtener el número

logistic growth growth of biological organisms which slows as resources are depleted

crecimiento logístico crecimiento de organismos biológicos que se desacelera cuando los recursos se agotan

lungs organs of the respiratory system that bring oxygen-rich air into the body and send oxygen-poor air out of the body

pulmones órganos del sistema respiratorio que traen aire rico en oxígeno al cuerpo y expulsan aire pobre en oxígeno fuera del cuerpo

M

macroevolution large-scale evolutionary change above the species level that leads to the development or loss of many new species over time

macroevolución cambio evolutivo a larga escala por encima del nivel de especie que conduce al desarrollo o pérdida de muchas especies nuevas con el paso del tiempo

magnetic field a set of lines that defines the motion of charged particles near a magnet

campo magnético conjunto de líneas que definen el movimiento de partículas cargadas cerca de un imán

mantle the layer of solid rock between Earth's crust and core

manto capa de roca sólida entre la corteza y el centro de la Tierra

marine biome a major ecological community defined by abundant water, coral reefs, and estuaries which supply most of the worlds oxygen supply

bioma marino gran comunidad ecológica definida por abundancia de agua, arrecifes de coral y estuarios; estos biomas proporcionan la mayor parte del oxígeno del mundo.

mass extinction the loss of many species throughout the world in a short period of time

extinción masiva pérdida de muchas especies en todo el mundo en un periodo de tiempo corto

meiosis a form of cell division that takes place in organisms that undergo sexual reproduction. Meiosis specifically results in the formation of four haploid reproductive cells (gametes).

meiosis forma de división celular que ocurre en organismos con reproducción sexual. La meiosis da como resultado específico la formación de cuatro células reproductivas haploides (gametos).

methane molecule comprising one carbon atom and four hydrogen atoms

mid-ocean ridge an oceanic rift zone that consists of long mountain chains with a central rift valley; divergent boundary

Milankovitch cycle periodic changes in Earth's climate-warming, followed by the onset of ice ages, followed by warming again; caused by irregularities in Earth's rotation and orbit

mineral a naturally occurring, inorganic solid with a definite chemical composition and characteristic crystalline structure

mitochondria an organelle in eukaryotic cells that is the site of cellular respiration and generates most of the cell's ATP

mitosis the process of cell division where one cell splits into two identical cells

molecular sequences the identity of the sequence of monomers in a polymer; the sequence of units that make up a larger molecule

monohybrid cross a cross between two organisms that differ with respect to a single trait

monosomy the condition of having a single copy of a given chromosome

mRNA messenger RNA; RNA transcribed from a protein coding gene that travels to the ribosome and is translated into a protein

multiple allele a gene that exists in three or more alleles in a population

metano molécula que tiene un átomo de carbono y cuatro átomos de hidrógeno

dorsales centro-oceánicas zona dorsal centro-oceánica que consta de largas cadenas montañosas con una fosa tectónica central; límite divergente

ciclo de Milankovitch cambios periódicos en el calentamiento del clima de la Tierra seguido por un comienzo de edades de hielo, seguidas por un nuevo calentamiento; producido por irregularidades en la rotación y en la órbita de la Tierra

mineral sólido inorgánico natural con una composición química definida y una estructura cristalina característica

mitocondria orgánulo de las células eucariotas donde tiene lugar la respiración celular y donde se genera la mayor parte del ATP de la célula

mitosis proceso de división celular en el cual la célula se divide en dos células idénticas

secuencias moleculares identidad de la secuencia de monómeros en un polímero; secuencia de unidades que constituye una molécula más grande

cruce monohíbrido cruce entre dos organismos que son diferentes respecto a un solo rasgo

monosomía condición de tener una sola copia de un cromosoma dado

ARNm ARN mensajero; ARN transcrito del código genético de una proteína que viaja al ribosoma y es traducido a proteína

alelo múltiple gen que existe en tres o más alelos en una población

GLOSSARY

muscle an organ of the muscular system: Muscles can be either voluntary, such as a biceps, or involuntary, such as heart muscle. (related word: muscular)

músculo órgano del sistema muscular: los músculos pueden ser voluntarios, como por ejemplo los bíceps, o involuntarios, como por ejemplo el músculo cardíaco (palabra relacionada: muscular)

muscular system the body system that permits movement and locomotion in animals

sistema muscular sistema corporal que permite el movimiento y la locomoción de los animales

mutation a change in the nucleotide sequence of an organism's genome; also a change in the amino acid sequence of a protein as a result of a mutation in the gene; natural selection can act upon a mutation that results in a change in phenotype

mutación cambio en la secuencia nucleotídica del genoma de un organismo; también un cambio en la secuencia del aminoácido de una proteína como resultado de una mutación en el gen; la selección natural puede actuar sobre una mutación que da como resultado un cambio en el fenotipo

mutualism a relationship between two species of a plant, animal, or fungus in which one lives off the other and both organisms benefit

mutualismo relación entre dos especies de una planta, animal u hongos en la cual uno vive a expensas del otro y ambos organismos obtienen beneficio

N

natural gas a nonrenewable fossil fuel that exists in the form of a gas; like all fossil fuels, natural gas formed over many millions of years as ancient, dead organisms were gradually compressed and heated deep beneath Earth's surface

gas natural combustible fósil no renovable que existe en forma de gas; al igual que todos los combustibles fósiles, el gas natural se formó a lo largo de muchos millones de años a partir de restos de antiguos organismos muertos que fueron sometidos a presión y calor en las profundidades de la Tierra

natural resource a mineral, organic material, or fuel deposit that is currently or may become available for human use

recurso natural depósito de mineral, materia orgánica o combustible que está disponible para el ser humano o puede estar disponible

natural selection the process by which traits or alleles become more or less frequent in a population, depending on the advantage or disadvantage they confer on the survival and reproduction of the organism

selección natural proceso mediante el cual los rasgos o alelos se hacen más o menos frecuentes en una población, dependiendo de las ventajas o desventajas que aporten para la supervivencia y la reproducción de los organismos

nerve a cell of the nervous system that carries signals to the body from the brain, and from the body to the brain and/or spinal cord

nervio célula del sistema nervioso que lleva señales al cuerpo desde el cerebro, y desde el cuerpo al cerebro y/o médula espinal

nervous system the system of the body that carries information to all parts of the body: The nervous system relies on nerve cells to move electrical signals to the body from the brain, and from the body to the brain and/or spinal cord.

sistema nervioso sistema del cuerpo que transporta información a todas las partes del cuerpo: El sistema nervioso depende de las células nerviosas para transportar señales eléctricas al cuerpo desde el cerebro y desde el cuerpo al cerebro y/o la médula espinal

neuron a cell in the nervous system responsible for sending neural messages

neurona célula del sistema nervioso encargada de enviar los mensajes neuronales

niche the unique physical environment occupied, and functions performed by, a species

nicho medio ambiente físico particular que ocupa y las funciones que realiza una especie

nitrogen cycle a process in which nitrogen in the atmosphere enters the soil and becomes part of living organisms then eventually returns to the atmosphere

ciclo del nitrógeno proceso en el cual el nitrógeno de la atmósfera penetra en el suelo y se convierte en parte de los organismos vivos, luego, con el tiempo, regresa a la atmósfera

nonconformity an unconformity in which sedimentary rock layers overlie an erosion surface cut into igneous or metamorphic rocks

inconformidad discordancia en la cual las capas de rocas sedimentaria están sobre una superficie de erosión cortada en rocas ígneas o metamórficas

nondisjunction an error in meiosis in which a pair of chromosomes does not separate correctly during meiosis I or meiosis II

no disyunción error en la meiosis en el cual un par de cromosomas no se separa correctamente durante la meiosis I o la meiosis II

non-point source origin of pollution that is not from a single location

fuente no puntual origen de contaminación que no surge de un solo lugar

nonrenewable resource a natural resource of which a finite amount exists, or one which cannot be replaced with currently available technologies

recurso no renovable recurso natural del cual existe una cantidad finita, o uno que no puede remplazarse con las tecnologías actualmente disponibles

nucleic acid a complex organic substance present in all living cells composed of a sugar, a base compound, and a phosphate group; includes both DNA and RNA

ácido nucleico sustancia orgánica compleja que está presente en todas las células vivas, compuesta por azúcar, un compuesto base y un grupo fosfato; en ellos se incluye el ADN y el ARN

nucleus (cell) an organelle in eukaryotic cells that contains the cell's chromosomes

núcleo (de la célula) orgánulo en las células eucariotas que contiene los cromosomas de la célula

nutrients an element or compound that an organism must consume or synthesize in order to survive

nutriente elemento o compuesto que un organismo debe consumir o sintetizar para sobrevivir

GLOSSARY

O

oceanic crust the portion of Earth's crust that makes up the ocean floor and is generally denser and thinner than continental crust

corteza oceánica parte de la corteza terrestre que conforma el fondo del océano y que por lo general es más densa y más fina que la corteza continental

oil a liquid fossil fuel composed primarily of carbon that forms from decomposed plant materials or algae

petróleo combustible fósil líquido compuesto principalmente por carbono que se forma a partir de materia vegetal o algas descompuestas

omnivore an animal that eats plants as well as other animals

omnívoro animal que se alimenta de plantas y de otros animales

oogenesis the process that results in the production of female gametes (eggs, or ova)

oogénesis proceso que da como resultado la producción de gametos femeninos (óvulos); también se denomina ovogénesis

organelle one of many membrane-enclosed structures that are found in the cytosol of eukaryotic cells, each with its own specific function

orgánulo una o varias estructuras encerradas en una membrana que se hallan en el citosol de las células eucariotas, cada uno de los cuales tiene una función específica

organic molecule a molecule found in or produced by living systems which contains carbon

molécula orgánica molécula que se halla en los sistemas vivos o que es producida por ellos y que contiene carbono

orogeny process of mountain formation by folding of Earth's crust

orogénesis proceso de formación de las montañas por plegamiento de la corteza de la Tierra

osmosis the diffusion of water through a selectively permeable membrane towards the side of the membrane with a higher solute concentration

ósmosis difusión de agua de manera selectiva a través de una membrana permeable hasta el otro lado de la membrana con una concentración de soluto más elevada

outer core the liquid outer portion of Earth's core, composed primarily of iron and nickel

núcleo externo parte exterior, líquida, del núcleo de la Tierra, compuesto principalmente por hierro y níquel

oxidation a chemical reaction resulting in the loss of electrons by a metal; for example, when iron rusts

oxidación reacción química que resulta de la pérdida de electrones por parte de un metal; por ejemplo, cuando el hierro se oxida

oxidizing agent the reactant that accepts electrons during an oxidation-reduction (redox) reaction

oxidante (agente oxidante) reactante que acepta electrones durante una reacción de reducción-oxidación (reacción redox)

ozone a molecule composed of three oxygen atoms (O^3)

ozono molécula compuesta por tres átomos de oxígeno (O^3)

P

paleontology the study of fossils and the fossil record

paleontología estudio de los fósiles y los registros fósiles

Pangaea the large supercontinent at the end of the Paleozoic Era consisting of all the land on Earth, including all seven continents and other landmasses

Pangea gran supercontinente al final de la Era Paleozoica que abarcaba toda la tierra de la Tierra, es decir, todos los siete continentes y las demás masas de tierra

parapatric speciation the evolution of an single population into more than one species as a result of individuals mating with their geographic neighbors rather than randomly within the population as a whole

especiación parapátrica evolución de una sola población en más de una especie como resultado de emparejamientos de los individuos con sus vecinos geográficos más bien que dentro de la misma población como un todo

parasitism a certain type of non-mutual relationship found between two different species in which one organism known as the parasite benefits at the expense of the other organism

parasitismo cierto tipo de relación no mutua que se produce entre dos especies diferentes en la que un organismo, conocido como parásito, se beneficia a expensas del otro organismo

passive transport movement of ions or molecules across a membrane down a concentration gradient ; it does not require the input of energy

transporte pasivo movimiento de iones o moléculas a través de una membrana hasta un gradiente de concentración; no requiere energía

per capita by individual persons or each individual

per cápita por persona o cada individuo

petroleum any form of naturally occurring hydrocarbons

petróleo hidrocarburo que se encuentra en la naturaleza en forma líquida

phenotype the observable traits of an individual, which are passed on from parent to offspring

fenotipo rasgo observable de un individuo que se pasa de un progenitor a su descendencia

phloem living tissue that transports nutrients to all parts of a plant

floema tejido vivo que transporta nutrientes a todas las partes de una planta

phospholipid bilayer (phospholipid membrane) a thin membrane made of two layers of phospholipids with the fatty acid chains in each layer facing towards the center of the membrane

bicapa fosfolipídica fina membrana compuesta por dos capas de fosfolípidos con las cadenas de ácidos grasos en cada capa mirando hacia el centro de la membrana

phosphorous cycle the transfer of phosphorous between the biosphere, lithosphere, and hydrosphere

ciclo del fósforo transferencia del fósforo entre la biosfera, litosfera e hidrosfera

GLOSSARY

photosynthesis the biological process by which most plants, some algae, and some bacteria produce organic compounds for their food from water and carbon dioxide using solar energy

fotosíntesis proceso biológico por el cual la mayoría de las plantas, algunas algas y algunas bacterias producen compuestos orgánicos que les sirven de alimento; estos compuestos los producen a partir de agua y dióxido de carbono usando la energía solar

phylogenetics the study of evolutionary relationships between species and other taxonomic groups

filogenética estudio de las relaciones evolutivas entre las especies y otros grupos taxonómicos

phylum a taxonomic category between class and kingdom

filo categoría taxonómica entre la clase y el reino

pioneer species species which colonize and inhabit land which has not yet been settled; typically leads to ecological succession

especie pionera especie que coloniza y ocupa una tierra que todavía no ha sido ocupada por otras; comúnmente conduce a una sucesión ecológica

plankton small organisms that drift through bodies of water; include animals, plants, and bacteria

plancton pequeños organismos que van a la deriva a través de los cuerpos de agua; incluye animales, plantas y bacterias

plant cell a form of a eukaryotic cell that is distinct from animal cells in that it possesses a rigid cell wall, large vacuole, and chloroplasts

célula vegetal tipo de célula eucariota que se diferencia de las células animales en que posee una pared celular rígida, una gran vacuola y cloroplastos

plate motion the motion of tectonic plates, which occurs at a rate of a few cent meters per year

movimiento de placas movimiento de las placas tectónicas que ocurre a un ritmo de pocos centímetros al año

plate tectonics the theory that describes the movement and recycling of segments of Earth's crust, called tectonic plates

tectónica de placas teoría que describe el movimiento y reciclaje de fragmentos de corteza terrestre, llamados placas tectónicas

point source origin of pollution from a single location

fuente puntual origen de la contaminación desde un solo lugar

polypeptide a chain of amino acid molecules, such as a protein

polipéptido cadena de moléculas de aminoácidos, como una proteína

population the group of organisms of the same species living in the same area

población grupo de organismos de la misma especie que viven en la misma área

population demographics statistical characteristics of a population

demografía de una población características estadísticas de una población

population density number of individuals per unit area	**densidad poblacional** número de individuos por área unitaria
potential energy the amount of energy that is stored in an object; energy that an object has because of its position relative to other objects	**energía potencial** cantidad de energía almacenada en un objeto; energía que tiene un objeto por su posición respecto a otros objetos
precipitation (crystallization) the settling of solid particles, such as crystals, to the bottom of an aqueous solution	**precipitación (cristalización)** asentamiento de partículas sólidas, como cristales en el fondo de una disolución acuosa
precipitation (weather) the falling of liquid or frozen water droplets from clouds	**precipitación (meteorología)** caída de agua líquida o congelada de las nubes
predation a certain type of relationship primarily found between two animal species in which one hunts, kills, and feeds off the other	**depredación** cierto tipo de relación que se produce principalmente entre dos especies animales en la cual una caza, mata y come a la otra como alimento
predator an organism, usually an animal, that kills another organism for food	**depredador** organismo, por lo general un animal, que mata a otro organismo para alimentarse
prey an organism that is hunted and eaten by another organism	**presa** organismo al que caza y come otro organismo
primary consumer organisms that consume producers for energy and nutrients	**consumidor primario** organismo que consume productores para obtener energía y nutrientes
primary succession one type of biological and ecological succession that involves the growth of plant life in a newly developed area defined by rock or other minerals and either no or very little soil	**sucesión primaria** tipo de sucesión biológica y ecológica que involucra el crecimiento de vida vegetal en un área de desarrollo reciente definida por rocas u otros minerales y ningún o muy poco suelo
principle of inclusions the scientific law stating that inclusions or fragments in a rock unit are older than the rock unit itself	**principio de inclusiones** ley científica que establece que las inclusiones o fragmentos dentro de una unidad rocosa son más antiguos que la propia unidad rocosa
principle of lateral continuity the scientific law stating that sedimentary layers extend horizontally outward in all directions until they terminate	**principio de la continuidad lateral** ley científica que establece que las capas sedimentarias se extienden horizontalmente hacia el exterior en todas direcciones hasta que terminan

GLOSSARY

principle of superposition   the scientific law stating that, in undisturbed rock layers, each layer is younger than the layer beneath it, and older than the layer above it

principio de superposición   ley científica que estable que, en capas de roca inalteradas, cada capa es más joven que la capa bajo ella y más antigua que la capa sobre ella

prokaryotic cell   a type of cell that is simple in structure and lacks a membrane-enclosed nucleus and membrane-enclosed organelles; they have an outer cell wall that gives them shape

célula procariota   tipo de célula de estructura simple que carece de un núcleo dentro de una membrana y orgánulos dentro de una membrana; tiene una pared celular externa que le da forma

protein   an organic molecule composed primarily of amino acids joined by peptide bonds in one or more chains; proteins function as enzymes, signaling molecules, structural molecules, and as a source of energy, among other functions

proteína   molécula orgánica compuesta principalmente por aminoácidos unidos por enlaces peptídicos en una o más cadenas; las proteínas funcionan como enzimas, moléculas señalizadoras, moléculas estructurales y fuentes de energía, entre otras funciones

R

radioactive decay   a process by which an unstable atom loses energy by emitting ionized particles over a period of time

desintegración radiactiva   proceso por el cual un átomo inestable pierde energía al emitir partículas ionizadas durante un periodo de tiempo

radioactive isotope   an isotope with an unstable nucleus, which can spontaneously decay

isótopo radiactivo   isótopo con un núcleo inestable, el cual puede desintegrarse espontáneamente

radioactivity   the spontaneous emission of radiation, which is the process in which unstable atoms break down into smaller atoms, releasing energy

radiactividad   emisión de radiación espontánea, que es el proceso en el cual un átomo inestable se divide en átomos más pequeños y libera energía

recessive allele   a genetic trait that lacks the ability to manifest itself when a dominant gene is present

alelo recesivo   rasgo genético que no tiene la capacidad de manifestarse cuando está presente un gen dominante

redox reaction   a short-hand term for an oxidation-reduction reaction, a chemical reaction that involves the transfer of one or more electrons from one species to another

reacción redox   término abreviado para reacción de reducción-oxidación; reacción química que implica la transferencia de uno o más electrones de una especie a otra

reducing agent   the reactant that donates electrons during an oxidation-reduction (redox) reaction

agente reductor   reactante que dona electrones durante una reacción de reducción-oxidación (reacción redox)

reduction a decrease in the oxidation state of an atom or molecule due to the gain of electrons

renewable resource a natural resource that can be replaced

replication fork Y-shaped structure that forms during the process of DNA replication; the unseparated double stranded DNA represents the base of the Y; the separated single strands are the arms of the Y

respiratory system the system of the body that brings oxygen into the body and releases carbon dioxide

ribosome molecular structure that facilitates DNA translation into protein

ridge push the sliding of oceanic lithosphere downward and away from a mid-ocean ridge due to the higher elevation of the mid-ocean ridge relative to a subduction zone

rift valley a tectonic valley that forms by extensional stress which causes fracturing and the formation of normal faults

rift zone a divergent boundary where the crust is pulled apart

RNA ribonucleic acid (RNA): one of the macromolecules that determines protein synthesis in the cell

RNA polymerase enzyme that transcribes RNA from a DNA template

rock cycle the process during which rocks are formed, change, wear down, and are formed again over long periods of time

rRNA ribosomal RNA; the RNA component of ribosomes, the site of protein synthesis.

reducción disminución en el estado de oxidación de un átomo o una molécula debido a la ganancia de electrones

recurso renovable recurso natural que puede remplazarse

horquilla de replicación estructuras en forma de Y que se forman durante el proceso de replicación del ADN; la doble hebra de ADN no separada representa la base de la Y; las hebras simples separadas forman los brazos de la Y

sistema respiratorio sistema del cuerpo que lleva oxígeno al cuerpo y libera dióxido de carbono

ribosoma estructura molecular que facilita la traducción del ADN en proteína

empuje de las dorsales deslizamiento de la litosfera oceánica hacia abajo y alejándose de una dorsal centro-oceánica debido a la mayor elevación de la dorsal centro-oceánica respecto a la zona de subducción

fosa tectónica valle tectónico que se forma por tensión de extensión, la cual produce la fractura y la formación de fallas normales

zona de fractura límite divergente en el cual la corteza terrestre se separa

ARN ácido ribonucleico (ARN): una de las macromoléculas que determinan la síntesis de las proteínas en las células

ARN-polimerasa enzima que transcribe el ARN de un molde de ADN

ciclo de las rocas proceso durante el cual las rocas se forman, cambian, se desgastan y se vuelven a formar a lo largo de grandes periodos de tiempo

ARNr ARN ribosómico; el componente ARN de los ribosomas, lugar de la síntesis de las proteínas

GLOSSARY

S

saccharide organic molecule including sugars, starch and cellulose

sacárido molécula orgánica que incluye azúcares, almidón y celulosa

scavenger organism that feeds on the remains of other organisms

carroñero organismo que se alimenta de los restos de otros organismos

seafloor spreading the process by which new oceanic lithosphere forms at mid-ocean ridges as tectonic plates pull away from each other

expansión del fondo oceánico proceso por el cual se forma nueva litosfera oceánica en las dorsales centro-oceánicas a medida que las placas tectónicas se separan una de otra

secondary consumer an animal which feeds on primary consumers in the food chain

consumidor secundario animal que se alimenta de consumidores primarios en la cadena alimentaria

secondary succession one type of biological and ecological succession that involves the growth of plant life in an area that previously saw growth, but was destroyed for any reason

sucesión secundaria tipo de sucesión biológica y ecológica que involucra el crecimiento de vida vegetal en un área en la que creció con anterioridad pero donde fue destruida por cualquier razón

sediment solid material, moved by wind, water, and other forces, that settles on the surface of land or the bottom of a body of water

sedimento material sólido transportado por el viento, el agua y otras fuerzas, que se asienta en la superficie de la tierra o en el fondo de un cuerpo de agua

selective permeability property of a membrane to allow passage of specific molecules

permeabilidad selectiva propiedad de una membrana que permite el paso de moléculas específicas

sex-linked disorders a disorder that is determined by an alteration in the number of the X or Y sex chromosomes or by a defective gene on a sex chromosome

enfermedades ligadas al sexo enfermedad o trastorno determinado por una alteración del número de cromosomas sexuales X o Y o por un gen defectuoso en un cromosoma sexual

sexual reproduction the production of offspring in which two parents give rise to offspring with combinations of genes inherited from both parents via union of the gametes

reproducción sexual producción de descendencia en la cual dos progenitores dan lugar a su descendencia con combinaciones de genes heredados de ambos progenitores mediante la unión de gametos

sexual selection a type of natural selection where selective pressures result from the selection of mates with particular characteristics

selección sexual tipo de selección natural en la que las presiones selectivas son el resultado de la selección de parejas con características particulares

single-gene disorders a disorder that is the result of a defect or mutation in one gene

skeletal system the network of solid materials that give an organism's body its structure

slab pull the gravitational force on dense oceanic lithosphere that forces one plate to slide beneath another

soil the fertile, outermost layer of Earth's crust; composed of bits of rocks and minerals mixed with decomposing plant and animal material

solar energy radiant energy that comes from the sun

solar wind a continuous stream of charged particles emitted by the sun which flows outward into the solar system

somatic cell nuclear transfer a laboratory technique for creating a clonal organism by combining an egg and a donor nucleus; can be used as the first step in the process of reproductive cloning

species a group of organisms that share similar characteristics and can mate with each other to produce offspring

spermatogenesis the process that results in the production of male gametes (sperm)

spontaneous mutation a change in DNA sequence that occurs during the process of DNA replication; an error in DNA replication

stabilizing selection a type of natural selection where selective pressures favor the middle of the phenotypic range of traits

trastornos de un solo gen trastorno que es el resultado de un defecto o mutación en un solo gen

sistema esquelético red de materiales sólidos que proporcionan al cuerpo de un organismo sus estructura

fuerza de tracción de la placa (o slab pull) fuerza gravitacional en la densa litosfera oceánica que hace que una placa se deslice bajo otra

suelo capa fértil y más exterior de la Tierra; compuesta por trozos de rocas y minerales mezclados con materia animal y vegetal en descomposición

energía solar energía radiante que procede del Sol

viento solar corriente continua de partículas cargadas que emite el Sol la cual fluye hacia afuera, dentro del sistema solar

transferencia nuclear de células somáticas técnica de laboratorio para clonar organismos combinando un huevo u óvulo y un núcleo de un donante; puede usarse como primer paso en el proceso de clonado reproductivo

especie grupo de organismos que comparten características similares y que se pueden aparear para producir descendencia

espermatogénesis proceso que da como resultado la producción de gametos masculinos (espermatozoides)

mutación espontánea cambio en la secuencia de ADN que tiene lugar durante el proceso de replicación del ADN; error en la replicación del ADN

selección estabilizadora tipo de selección natural en la que las presiones selectivas favorecen una zona media del rango fenotípico de rasgos

Glossary

GLOSSARY

starch a long-chain carbohydrate formed from glucose units joined by glycosidic bonds

almidón larga cadena de hidrato de carbono formada por unidades de glucosa unidas mediante enlaces glucosídicos

stem the part of a plant that grows away from the roots; supports leaves and flowers

tallo parte de un planta que crece en dirección contraria a las raíces; da soporte a hojas y flores

steroid organic compound with a four ring core structure

esteroide compuesto orgánico con una estructura nuclear de cuatro anillos

sterol a type of steroid with a hydroxyl substituent

esterol tipo de esteroide con un hidroxilo sustituyente

stroma the fluid portion of the chloroplast that surrounds the grana

estroma parte líquida del cloroplasto que rodea la grana

subduction the sinking of an oceanic plate beneath a plate of lesser density at a convergent boundary

subducción hundimiento de una placa oceánica bajo una placa de menor densidad en un límite convergente

subduction zone a convergent boundary where oceanic lithosphere is forced down into the asthenosphere under the lithosphere that comprises another, less dense tectonic plate

zona de subducción límite convergente donde la litosfera oceánica es obligada a descender al interior de la astenosfera, bajo la litosfera, que comprende otras placas tectónicas menos densas

substrate substance involved in chemical reaction

sustrato sustancia implicada en una reacción química

substrate (biochemistry) molecule acted upon by an enzyme

sustrato (bioquímica) molécula sobre la que actúa una enzima

succession the sequence of communities that develop in an area.

sucesión secuencia de comunidades que se desarrollan en un área

sucrose a disaccharide composed of glucose and fructose

sacarosa disacárido compuesto por glucosa y fructosa

sunspot a cooler darker spot on the surface of the sun

mancha solar mancha más oscura y fría sobre la superficie del Sol

superposition the ordering of sedimentary layers of rock with the oldest on the bottom and the youngest on top

superposición orden de las capas sedimentarias de roca, la más antigua se encuentra en el fondo y la más joven en la parte superior

survivorship curve   a grapheither Type I, II, or IIIthat depicts either the proportion or number of people surviving at each age for a specific group or species

curva de supervivencia   gráfica de tipo I, II o III que muestra la proporción o el número de personas que sobreviven a cada edad para un grupo específico de especies

suspension   a heterogeneous mixture in which moderate sized particles are suspended, not dissolved, in a liquid or gas where they are supported by buoyancy

suspensión   mezcla heterogénea en la cual unas finas partículas se encuentran en suspensión, no disueltas, en un líquido o un gas en el que son soportadas por la flotabilidad

sustainable   describes a material or resource that is able to meet the demands of current use and yet be maintained in usable quantities to meet indefinite future demands.

sostenible   describe un material o recurso que es capaz de satisfacer las demandas de uso actual y mantenerse en cantidades utilizables para satisfacer demandas futuras indefinidas

sympatric speciation   the evolution of a population into a new species when two populations of the same species favor different niches in an area but mate only within their niche

especiación simpátrica   evolución de una población en nuevas especies cuando dos poblaciones de la misma especie favorecen nichos diferentes en un área pero se emparejan solo dentro de su nicho

system   a related set of components that react with one another that may or may not interact with the surrounding area

sistema   conjunto de componentes relacionados que reaccionan unos con otros y que pueden o no interactuar con el entorno

T

taiga biome   an ecological community that is characterized by cold weather, coniferous trees, and few food sources in winter

bioma de taiga   comunidad ecológica que se caracteriza por un clima frío, coníferas y pocas fuentes de alimentación en el invierno

tar pit   an accumulation of a natural, black, highly viscous mixture of hydrocarbons (called bitumen) exposed at the land surface

pozo de brea   acumulación de una mezcla natural, negra y muy viscosa de hidrocarburos (llamada bitumen) expuesta en la superficie de la tierra

taxonomic level   one of eight levels used by scientists to classify organisms; each species, the lowest taxonomic level, can also be classified into one of the higher levels: genus, family, order, class, phylum kingdom, domain

nivel taxonómico   uno de los ocho niveles que usan los científicos para clasificar organismos; cada especie, el nivel taxonómico más bajo, puede clasificarse a su vez en uno de los niveles superiores: género, familia, orden, clase, filo, reino, dominio

Glossary | 379

GLOSSARY

taxonomic unit   a group of organisms that are evolutionarily related, having a common ancestor; a group of organisms that all belong to the same taxonomic level (the same phyla, genus, or other taxonomic group)

technology   the use of scientific knowledge to solve problems and the devices created by this process

temperate deciduous forest biome   an ecological community that is defined by five different zones, four distinct seasons, and a mixed climate with mixed precipitation

temperature   a measure of the average kinetic energy of the atoms in a system, used to express thermal energy in degrees

tertiary consumer   a third-level consumer that feeds only on secondary consumers

The Theory of Evolution   A theory that the various species of living organisms have their origin in common ancestors and that the distinguishable differences are due to heritable modifications in successive generations

thermal energy   Thermal energy is the movement of molecules that make up the object. The molecules move faster when heated

thermodynamics   the study of effects of changes in temperature, pressure, and volume using statistics to analyze the collective motion of their particles

thylakoids   flat membrane-enclosed structures inside chloroplasts that contain chlorophyll and other pigments; they are the site for the light dependent reactions of photosynthesis

unidad taxonómica   grupo de organismos con una relación evolutiva, que tiene un antepasado común; grupo de organismos que pertenecen al mismo nivel taxonómico (el mismo filo, género u otro grupo taxonómico)

tecnología   uso del conocimiento científico para resolver problemas y los dispositivos que se crean por ese proceso

bioma de bosque templado caducifolio   comunidad ecológica que se define por cinco zonas diferentes, cuatro estaciones distintas y climas mixtos con precipitaciones mixtas

temperatura   medida del porcentaje de energía cinética de los átomos de un sistema, se usa para expresar la energía térmica en grados

consumidor terciario   consumidor de tercer nivel que se alimenta solo de consumidores secundarios

teoría de la evolución   teoría que establece que varias especies de organismos vivos tienen su origen en antepasados comunes y que' las diferencias que se distinguen se deben a modificaciones heredadas en generaciones sucesivas

energía térmica   La energía térmica es el movimiento de las moléculas que conforman un objeto. Las moléculas se mueven más rápido cuando se calientan.

termodinámica   estudio de los efectos de los cambios en temperatura, presión y volumen usando la estadística para analizar el movimiento colectivo de las partículas

tilacoide   membrana plana que forma parte de las estructuras internas de los cloroplastos que contiene clorofila y otros pigmentos; es el lugar donde se llevan a cabo las reacciones de fotosíntesis que dependen de la luz

thymine one of the four nitrogenous bases contained in a DNA molecule; thymine is complementary to adenine	**timina** una de las cuatro bases nitrogenadas que forman parte de una molécula de ADN; la timina es complementaria a la adenina
tidal energy a form of hydroelectric energy which converts the energy of ocean tides into electricity	**energía mareomotriz** forma de energía hidroeléctrica que convierte la energía de las mareas oceánicas en electricidad
topography the physical features which define the relief of a landscape, such as mountains, valleys, and the shapes of landforms	**topografía** características físicas que definen el relieve de un lugar, como montañas, valles y la forma de los accidentes geográficos
topsoil the organic-rich, dark-colored soil on the surface of an area, defined as the A horizon	**mantillo** suelo de color oscuro y rico en elementos orgánicos que se halla en la superficie de un área, definido como horizonte A
trace fossil fossilized evidence of plant existence or animal movements such as root channels, footprints, and burrows	**traza fósil** evidencia fosilizada de la existencia de plantas o movimientos de animales, como canales de raíces, huellas de pies y madrigueras
transcription the process of synthesizing RNA from a DNA template	**transcripción** proceso de síntesis del ARN de un molde de ADN
transform boundary a tectonic plate boundary along which plates slide horizontally past one another in opposite directions	**límite transformante** límite de placas tectónicas a lo largo de la cual las placas se deslizan horizontalmente una junto a otra en direcciones opuestas
translation the process of building a protein based on a RNA template	**traducción** proceso por el cual se construye una proteína basada en un molde de ARN
transport the movement of materials	**trasporte** movimiento de materiales
tree ring a concentric layer of wood that a tree adds to its trunk and branches in one year; each successive ring represents a year of the tree's life	**anillo de árbol** capa de madera concéntrica que un árbol añade a su tronco y a sus ramas en un año; cada anillo sucesivo representa un año de la vida del árbol
trisomy a type of aneuploidy in which there are three copies of a particular chromosome	**trisomía** tipo de aneuploidía en la cual hay tres copias de un cromosoma particular

GLOSSARY

tRNA transfer ribonucleic acid; tRNA transfers amino acids to a growing protein chain during protein synthesis; different tRNA molecules have different anticodons and carry amino acids specific to their anticodon

ARNt ácido ribonucleico de transferencia; el ARNt transfiere aminoácidos a una cadena de proteínas en crecimiento durante la síntesis de proteínas; diferentes moléculas de ARNt tienen diferentes anticodones y llevan aminoácidos específicos a su anticodón

trophic level the position an organism occupies at each level of the food chain

nivel trófico posición que ocupa un organismo en cada nivel de la cadena alimentaria

tropical biome an ecological community that is defined by year-round warmth and significant rainfall

bioma tropical comunidad ecológica definida por temperaturas cálidas y lluvias significativas a lo largo de todo el año

tundra biome an ecological community that is defined by stark, treeless land, and very cold temperatures

bioma de tundra comunidad ecológica definida por tierras inhóspitas y sin árboles y temperaturas muy frías

U

unconformity a gap (or break) in the rock record where younger rocks are separated from older rocks with intervening periods of deposition completely missing

discordancia interrupción en el registro de la roca en la cual las rocas más jóvenes están separadas de las más antiguas con periodos intermedios de deposición completamente perdidos

uracil a nitrogenous (nitrogen containing) base found in RNA

uracilo base nitrogenada (que contiene nitrógeno) que se halla en el ARN

V

vein a blood vessel that moves blood towards the heart

vena vaso sanguíneo que transporta la sangre hacia el corazón

vertebrate an animal with a backbone

vertebrado animal con columna vertebral

W

water a molecule that contains two hydrogen atoms and one oxygen atom; often called "the universal solvent"

agua molécula que contiene dos átomos de hidrógeno y un átomo de oxígeno; con frecuencia se le llama "el solvente universal"

water cycle the continual movement of water between the land, ocean, and the air through predictable physical processes

ciclo del agua movimiento continuo del agua entre la tierra, el océano y el aire mediante procesos físicos predecibles

weathering the physical or chemical breakdown of rocks and minerals into smaller pieces or aqueous solutions on Earth's surface

meteorización desintegración física o química de rocas y minerales en trozos más pequeños o en soluciones acuosas en la superficie de la Tierra

weathering (physical) the breaking down of rock into smaller pieces by the action of wind, rain, and temperature change

meteorización (física) desintegración de las rocas en pequeños trozos por la acción del viento, de la lluvia y del cambio de temperatura

wind energy electricity generated by turbines rotated by wind

energía eólica electricidad generada por turbinas que hace girar el viento

X

xylem plant tissue that transports water upwards from the roots and also transports some nutrients

xilema tejido vegetal que transporta agua hacia arriba, desde las raíces, también transporta algunos nutrientes